Social Physics

ALSO BY ALEX PENTLAND

Honest Signals: How They Shape Our World

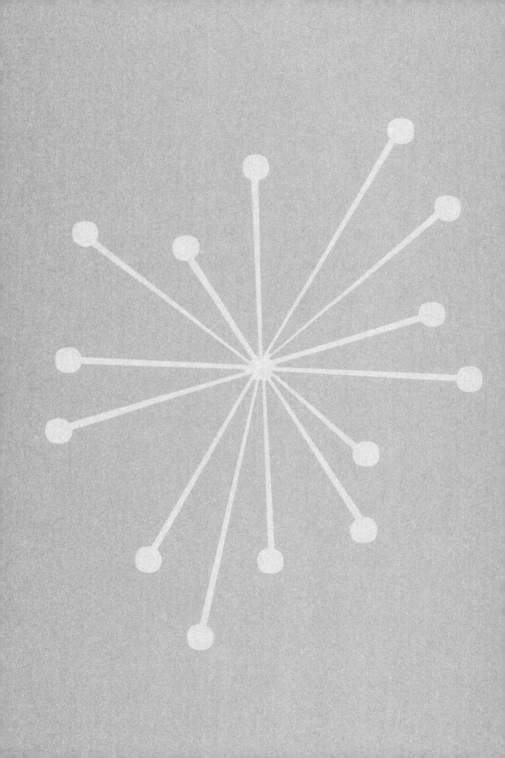

Social Physics

How Good Ideas Spread—
The Lessons from a New Science

ALEX PENTLAND

•

THE PENGUIN PRESS
New York
2014

THE PENGUIN PRESS
Published by the Penguin Group
Penguin Group (USA) LLC
375 Hudson Street
New York, New York 10014

USA • Canada • UK • Ireland • Australia
New Zealand • India • South Africa • China
penguin.com
A Penguin Random House Company

First published by The Penguin Press, a member of Penguin Group (USA) LLC, 2014

LIBRARY OF CONGRESS CATALOGING-IN-PUBLICATION DATA
Pentland, Alex, 1952–
Social physics : how good ideas spread—the lessons from a new science / Alex Pentland.
pages cm
Includes bibliographical references and index.
ISBN 978-1-59420-565-1
1. Inventions—Social aspects.   2. Technology transfer.   3. Science—Social aspects.
4. Social interaction.   I. Title.
T14.5.P45 2014
303.48'3—dc23   2013039929

Printed in the United States of America
1   3   5   7   9   10   8   6   4   2

DESIGNED BY AMANDA DEWEY

Preface

The Origin of the Book

I live in the future. The Massachusetts Institute of Technology (MIT), where I work, is at the center of the innovation universe; virtually any new idea or technology in the world travels through MIT before it shows up on the world stage. MIT is also part of the densest collection of start-up companies in the world (although Silicon Valley is larger). Moreover, the MIT Media Lab, my intellectual home, is probably the world's foremost place to *live* the future. For instance, fifteen years ago I ran the world's first cyborg collective, in which everyone lived and worked with wirelessly connected computers on their bodies and computer displays in their glasses. Many of these ideas eventually find their way out into the world; my former students now lead cutting-edge commercial projects such as Google Glass (glasses with computer screens built in) and Google+ (the world's second-largest social network).

My privileged position has given me a unique opportunity to

see firsthand how creative cultures can harvest new ideas, help them survive and grow, and finally, turn them into practical reality. Perhaps more important, I have also gotten to see how creative cultures must change in order to thrive in the hyperconnected, warp-speed world that is MIT, an environment that the rest of the world is now entering.

What I have learned from these experiences is that many of the traditional ideas we have about ourselves and how society works are wrong. It is not simply the brightest who have the best ideas; it is those who are best at harvesting ideas from others. It is not only the most determined who drive change; it is those who most fully engage with like-minded people. And it is not wealth or prestige that best motivates people; it is respect and help from peers.

These ideas are central to the success of the Media Lab, my research group, and to the entrepreneurship program that I direct. I don't teach traditional classes; instead, I bring in visitors with new ideas and get people to interact with others who are on the same journey. When I was academic head of the Media Lab I pushed to get rid of traditional grading; instead, we have tried to grow a community of peers where respect and collaboration on real-world projects is the currency of success and further opportunity. We live in social networks, not in the classroom or laboratory.

The origins of this book are a sharp clash of cultures between how I do things at the Media Lab and how things are done elsewhere in the world. For example, when I created the Media Lab Asia as a distributed organization across several universities in India, one of the biggest problems I encountered was that researchers at each university were isolated from one another and therefore their research was stagnant and unproductive. People working in the same field, and sometimes even at the same university, had literally never met each other because the university administrators and the funding agencies thought it was sufficient to have the researchers read each other's papers and that they didn't need to travel to meetings or conferences. It was only when they began to meet and spend *informal* time together that new ideas began to bubble up and new ways of approaching problems began to spread.

I have seen the same lack of understanding in many senior government leaders and CEOs of multinational companies in my role at the World Economic Forum, where I have co-led "hyperconnected world" discussions that seek solutions to the challenges posed by big data and specifically the uncontrolled spread of private personal information. It has become clear to me that there is a huge difference between the way most world leaders and CEOs think about innovation and collective action and the examples I see from my perch at MIT. Most people think in relatively static terms, such as competition, rules, and (sometimes) complexity. I think in more dynamic, evolutionary terms, paying attention to the flow of ideas within networks, the creation of social norms, and the processes that generate complexity. Most people think about using a framework centered on the individual and the eventual steady-state outcome, whereas I think in terms of social physics: growth processes within networks.

To understand this difference in thinking, I began a decade-long research program to develop a rigorous intellectual framework that extends current individual-centric economic and policy thinking by including social interactions. It posits social learning and social pressure as primary forces that drive the evolution of culture and govern much of the hyperconnected world. This research program has been surprisingly successful from an academic perspective, with each part of the social physics framework being mapped out in papers published in the world's most selective scientific journals. My expectation is that these papers will provide additional depth to the fields of complexity and network science, as well as provide a new view on evolutionary dynamics.

But as we all know, academic papers are, well, academic. So I've also helped move these ideas out into the real world, creating half a dozen start-up companies that use them to help firms become more productive and creative, to make the mobile social Web smarter, to make it possible for the average person to be a successful investor, and to help support the social and mental health of our society. Again, these real-world endeavors have been surprisingly successful, in no small part because of the talented and visionary former students who have become the CEOs of these companies.

This book marks the launch of a larger discussion. The goal is to get the language of social physics into general use, where it can provide much needed nuance to the traditional language of market competition and regulation. In a world of hyperconnectivity, where social dynamics are such an important determinant of outcomes, a better understanding of social physics has become critical.

Acknowledgments

It is with deepest appreciation that I would like to acknowledge Tracy Heibeck and her work throughout every step of this book. She made sure each argument was carefully untangled and every phrase was properly turned, and but for the vagaries of the publishing world, she would be listed as coauthor; Max Brockman and Scott Moyers, whose enthusiasm first made this book possible, and then ensured that it was not only readable but even exciting; Mally Anderson, for her thoughtful editing; and last but not at all least, to my students, postdocs, and colleagues, whose research efforts helped forge the ideas, experiments, techniques, and conclusions presented in this book.

Contents

PART III

Data-Driven Cities

9 · CITY SCIENCE

How Social Physics and Big Data Are Revolutionizing Our Understanding of Cities and Development   155

PART IV

Data-Driven Society

10 · DATA-DRIVEN SOCIETIES

What Will a Data-Driven Future Look Like?   177

Social Physics

From Ideas to Actions

USING BIG DATA TO UNDERSTAND HOW HUMAN SOCIETIES EVOLVE

Where do new ideas come from? How do they get put into action? How can we create social structures that are cooperative, productive, and creative? These are perhaps the most critical questions for any society, and they are especially important right now because of global competition, environmental challenges, and the failure of governments to act.

In the past few centuries we have seen Western culture thrive, in large part because of the paradigms inherited from Enlightenment thinkers such as Adam Smith and John Locke. Their intellectual frameworks offered answers to these critical questions. From that base we created a pluralistic society in which both the distribution of goods and the policies of government are determined by competition and bargaining. Our open civil society outcompeted more top-down, centralized societies, and now free markets and political elections are being experimented with in almost every country around the world.

In the last few years, however, our lives have been transformed by networks that combine people and computers, allowing much greater participation and much faster change. As the Internet makes our lives increasingly connected, events seem to move faster and faster. We are drowning in information, so much so that we don't know what items to pay attention to and which to ignore.

As a result, our world sometimes seems to be on the edge of spinning out of control, with posts on social media such as Twitter causing stock market crashes and overthrowing governments. So even though the use of digital networks has already converted the workings of our economy, business, government, and politics, we still don't fully understand the fundamental essence of these new human-machine networks. Suddenly our society has become a combination of humans and technology that has powers and weaknesses different from any we have ever lived in before.

Unfortunately, we don't really know what to do about it. Our ways of understanding and managing the world were forged in a statelier, less connected time. Our current conception of society was born in the late 1700s during the Enlightenment and crystallized into its current form during the first half of the twentieth century. Things moved more slowly back then, and usually it was only a small group of traders, politicians, or wealthy families who really moved things along. Therefore, when we think about how to manage our society, we speak of "markets" and "political classes," abstractions that events move slowly, so everyone has pretty much the same information and so people have time to act rationally.

In today's light-speed, hyperconnected world, these assumptions are being stretched past the breaking point. Today virtual crowds can form in minutes and often consist of millions of people from all over the world—and with each new day it may be a different set of millions of people contributing and commenting. We are no longer in the era of financial exchanges with physical trading floors and political conventions with smoke-filled back rooms, where small groups of people haggle until they come to mutually acceptable deals.

To understand our new world we must extend familiar economic and political ideas to include the effects of these millions of people learning from each other and influencing each other's opinions. We can no longer think of ourselves as only individuals reaching carefully considered decisions; we must include the dynamic social effects that influence our individual decisions and drive economic bubbles, political revolutions, and the Internet economy.

Adam Smith himself understood that it is our social fabric that guides the "invisible hand" of the market and not just competition alone. In his book *Theory of Moral Sentiments* he argued that it was human nature to exchange not only goods but also ideas, assistance, and favors out of sympathy.[1] Furthermore, he thought that these social exchanges guided capitalism to create solutions for the good of the community. Smith, though, lived in an era where almost all the bourgeois residents in a city knew each other and were constrained by social pressure to be good citizens. Without the obligations provided by strong social ties, capitalism often turns rapacious and politics turn poisonous. In our new hyperconnected world, most ties are weak, and all too often the invisible hand no longer functions.

The goal of this book is to develop a social physics that extends economic and political thinking by including not only competitive forces but also exchanges of ideas, information, social pressure, and social status in order to more fully explain human behavior. To accomplish this we will have to explain not only how social interactions affect individual goals and decisions but, more important, how these social effects produce Adam Smith's otherwise mysterious invisible hand.[2] Only once we understand how social interactions work together with competitive forces can we hope to ensure stability and fairness in our hyperconnected, networked society.

What Is Social Physics?

Social physics is a quantitative social science that describes reliable, mathematical connections between information and idea flow on the one hand and people's behavior on the other. Social physics helps us understand how ideas flow from person to person through the mechanism of social learning and how this flow of ideas ends up shaping the norms, productivity, and creative output of our companies, cities, and societies. It enables us to predict the productivity of small groups, of departments within companies, and even of entire cities. It also helps us tune communication networks so that we can reliably make better decisions and become more productive.

The key insights obtained with social physics all have to do with the flow of ideas between people. This flow of ideas can be seen in the pattern of telephone calls or social media messaging, of course, but also by assessing how much time people spend together and whether they go to the same places and have similar experiences. As we will see, flows of ideas are central to understanding society not only because timely information is critical to efficient systems but, more important, because the spread and combination of new ideas is what drives behavior change and innovation.

This focus on the flow of ideas is why I chose the name "social physics." Just as the goal of traditional physics is to understand how the flow of energy translates into changes in motion, social physics seeks to understand how the flow of ideas and information translates into changes in behavior.

As an example of social physics in action, consider the behavior of financial day traders who share tips on a social network. There are times when very few traders make much of a profit, results that are bad both for the traders and their brokers, who lose their business when they quit. To improve the traders' results, brokers have tried standard solutions, such as trying to improve the traders' knowledge and expertise. And these traditional remedies do have some effect: In one case, the performance of a group of traders increased by roughly 2 percent.

But then one broker agreed to let my MIT research laboratory try a social physics approach, drawing on our mathematical models of how ideas spread through social networks. By analyzing the millions of detailed messages among traders on a social network, we discovered that the effects of social influence within the network were too strong, causing the phenomenon of herding, in which the traders overreacted to each other, and so all tended to adopt the same trading strategy.

The mathematics of social physics indicated that the best approach to fixing this problem was to change the social network, in order to slow down the spread of new strategies within it. When we implemented these changes, it doubled the average return on investment, leaving the standard economic approaches in the dust.

Slowing down the spread of ideas is not something that is found in a standard management handbook. And this result was no accident, because we had mathematical analyses based on millions of bits of data that made it possible for us to devise precise interventions and predict precisely what the outcome would be. Those equations are part of the mathematics of social physics, as I will begin to explain in Chapter 2.

A Practical Science

The name social physics has a long history. Its original use was in the early 1800s, when, using an analogy derived from Newtonian physics, society was conceptualized as a vast machine. But society just isn't very machinelike. In the mid-twentieth century there was a second wave of interest in social physics, when it was discovered that many social indicators had statistical regularities such as the Zipf distribution[3] and the gravity law.[4] In parallel, social scientists have refined theories about the basic mechanisms of social interactions.[5] Most recently we have seen a wave of "sociophysics," through which we've discovered statistical regularities within human movement and communication, and interesting correlations with economic indicators.[6] As a consequence of these new types of data, social science theories have become much more quantitative.[7]

None of these efforts, though, really gets at the mechanism that drives societal changes and is the cause of these statistical regularities. Both theory and mathematical description remain fragmented and difficult to apply to practical problems. We need to move beyond merely describing social phenomena to building a causal theory of social structure. Progress in developing this represents steps toward what David Marr called a computational theory of behavior: a mathematical explanation of why society reacts as it does and how these reactions may (or may not) solve human problems.[8]

This sort of computational theory of behavior, which focuses on the human generative process, is what is required to build better social systems. Such a theory could tie together mechanisms of social interactions with our newly acquired massive amounts of behavior data in order to engineer better social systems.

This book presents the beginnings of such a practical theory, and is based on a series of my papers that have recently appeared in the world's leading scientific journals. This theory is a deceptively simple family of mathematical models that can be explained in plain English and gives a reasonably accurate account for the dozens of real-world examples described in this book. These examples include: financial decision making (including phenomena such as bubbles); "tipping point"–style cascades of behavior change, recruiting millions of people to help in a search, to save energy, or to get out and vote; as well as social influence and its role in shaping political views, purchasing behavior, and health choices.

The ultimate test of a practical theory, of course, is whether or not it can be used to shape outcomes. Is it good enough for engineering? To answer this question I will show how this new theory is already being used to create better companies, cities, and social institutions. Almost uniquely among the social sciences, this new social physics framework provides quantitative results at scales ranging from small groups, to companies, to cities, and even to entire societies. Currently, a social physics framework is in daily use in several commercial deployments, serving tens of millions of people in tasks such as financial investing, health monitoring, marketing, improving company productivity, and boosting creative output.

In the end, though, the importance of a science of social physics will not be only its utility in providing accurate, useful mathematical predictions. If social physics is just complex mathematics, then its use will be restricted to specially trained experts. I believe that its ultimate impact also depends upon whether it provides people—e.g., government and industry leaders, academics, and average citizens—a language that is better than the old vocabulary of markets and classes, capital and production. Words such as "markets," "political classes," and "social movements" shape our thinking about the world. They are useful, of course, but they also represent overly simplistic thinking; they therefore limit our ability to think clearly and effectively. In this book I will put forward a new set of concepts with which I believe we can more accurately discuss our world and plan the future.

Big Data

The engine that drives social physics is big data: the newly ubiquitous digital data now available about all aspects of human life. Social physics functions by analyzing patterns of human experience and idea exchange within the digital bread crumbs we all leave behind us as we move through the world—call records, credit card transactions, and GPS location fixes, among others. These data tell the story of everyday life by recording what each of us has chosen to do. And this is very different from what is put on Facebook; postings on Facebook are what people choose to tell each other, edited according to the standards of the day. Who we actually are is more accurately determined by where we spend our time and which things we buy, not just by what we say we do.[9]

The process of analyzing the patterns within these digital bread crumbs is called reality mining, and through it we can tell an enormous amount about who individuals are. My students and I have found that we can use it to tell if people are likely to get diabetes or whether someone is the sort of person who will pay back loans. And by analyzing these patterns across many people, we are discovering that we can begin to explain many things—crashes, revolutions, bubbles—that previously appeared to be random "acts of God." For this reason the magazine MIT *Technology Review* named our development of reality mining as one of the ten technologies that will change the world (for additional detail see the Reality Mining appendix).

The scientific method used in social physics is different from that used in most social sciences, because it principally relies on "living laboratories." What is a living lab? Let us imagine the ability to place an imaging chamber around an entire community and then to record and display every facet and dimension of behavior, communication, and social interaction among its members. Now think about doing this for up to several years while the members of the community go about their everyday lives. That is a living lab.

During the past decade, my students and I have developed the ability to build and deploy such living labs, measuring entire social organisms—groups, companies, and whole communities—on a second-by-second basis for up to years at a time. The method is simple: Measurements are made by collecting digital bread crumbs from the sensors from cell phones, postings on social media, purchases with credit cards, and more.

To accomplish this I have developed legal and software tools to protect the rights and privacy of the people in these labs to insure that they are fully informed about what is happening to their data and that they maintain the right to opt out at any time. As I will explain, the solutions I have developed have been important in helping to improve the privacy protections of citizens around the world. (Details about these legal and software tools are contained in the Reality Mining and the openPDS appendices.)

All of those billions of telephone call records, credit card transactions, and GPS location fixes have provided scientists with a new lens that lets us examine society in fine-grained detail.[10] Just as when Dutch lens makers created the first practical lenses and thus enabled researchers to build the first microscopes and telescopes, my research lab and I have created tools that collect all the digital bread crumbs from an entire community, enabling us to build some of the first practical "socioscopes." These new tools give a view of life in all its complexity—and are the future of social science. Just as the microscope and telescope revolutionized the study of biology and astronomy, socioscopes in living labs will revolutionize the study of human behavior.

A Rich Social Science

Most current social science is based on either analysis of laboratory phenomena or on surveys—that is, on descriptions of averages or stereotypes. These approaches don't account for the complexity of real life, when all of our mental quirks operate at the same time. They also miss the critical fact that the details about the people we interact with, and how we interact with them, matter as much as market forces or class structures. Social phenomena are really made up of billions of small transactions between individuals— people trading not only goods and money but also information, ideas, or just gossip. There are patterns in those individual transactions that drive phenomena such as financial crashes and Arab springs. We need to understand these micropatterns because they don't just average out to the classical way of understanding society. Big data give us a chance to view society in all its complexity, through the millions of networks of person-to-person exchanges.

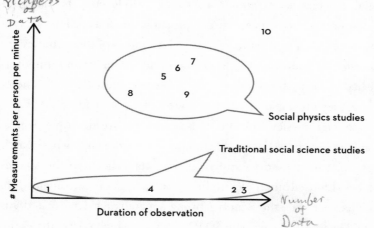

richness of Data

Number of Data

Figure 1. Qualitative overview of social science observatories and experiments, with the horizontal axis showing data collection duration and the vertical axis showing richness of the information collected. Data sets include: (1) Most social science experiments, (2) Midwest Field Station,[*] (3) Framingham Heart Study,[†] (4) Large Call Record data sets[‡] (5) Reality Mining,[§] (6) Social Evolution,[**] (7) Friends and Family,[††] (8) Sociometric Badge studies,[‡‡] (9) Data for Development (D4D) data set,[§§] (10) where the world is headed.

If we had a "god's eye," an all-seeing view, then we could potentially arrive at a true understanding of how society works and take steps to fix our problems. Unfortunately, as illustrated in Figure 1, almost all data from traditional social science (labeled 1) are near the (0,0) coordinate, meaning these data sets represent information

---

[*] Barker 1968.

[†] Dawber 1980.

[‡] Gonzalez et al. 2008; Eagle et al. 2010; Hidalgo and Rodriquez-Sickert 2008.

[§] Eagle and Pentland 2006.

[**] Madan et al. 2012.

[††] Aharony et al. 2011.

[‡‡] Pentland 2012b.

[§§] http://www.d4d.orange.com/home.

gathered from fewer than a hundred people, and for only a few hours. The studies labeled 2 and 3 are some of the largest social science studies to date.[11] Over the past decade, computational social scientists have begun to discover how to leverage big data and have been using data sets from companies such as cell phone carriers and social media firms. Typical examples of these studies are labeled 4. Unfortunately, even these large data sets are impoverished because they measure only a few variables at a time. Therefore, they provide a very limited view of human nature.

Social physics seeks the richest quantitative descriptions possible. The studies labeled 5, 6, and 7 are from my own research group, which used smartphones for data collection; the studies labeled 8 used smart electronic name badges, or "sociometers," to collect data (see the Reality Mining appendix for details); and 9 is the Data for Development (D4D) data set, which covers the entire population of the Ivory Coast.

Just a brief examination of Figure 1 makes it easy to see that these social physics data sets are many orders of magnitude richer than previous social science data sets. These large digital data sets contain extraordinary amounts of objective, continuous, and dense data that enable us to build quantitative, predictive models of human behavior in complex, everyday situations.

Importantly, the point labeled 10 is where the world is headed. In just a few short years we are likely to have incredibly rich data available about the behavior of virtually all of humanity—on a continuous basis. The data mostly already exist in cell phone networks, credit card databases, and elsewhere, but currently only technical gurus have access to it. As they become more widely available for scientific inquiry, however, the new science of social physics will gain further momentum. And once we develop a more precise visualization of the patterns of human life, we can hope to understand and manage our modern society in ways better suited to our complex, interconnected network of humans and technology.

In support of this book, I have placed several of the world's largest and most detailed living lab data sets onto the Web. These new digital information sources let us precisely measure patterns of interaction between people or between people and merchants, and to chart the patterns of experiences people have as they go about their lives. These living lab data sets comprise:

> *Friends and Family:* Roughly eighteen months of data from a small community of young families, with a wide variety of sociometric variables, including location, proximity, communication, purchasing, social media use, mobile apps, and sleep.[12] We measured thirty behavior variables every six minutes.[13] This study contains a total of 1.5 million hours of quantitative observation of the human social experience.

> *Social Evolution:* Nine months of data from a university dormitory, with location, proximity, and communication information measured every five minutes, along with health, political, and sociometric variables.[14] This study includes a total of 500,000 hours of quantitative observation.

> *Reality Mining:* Nine months of data from graduate students at two university laboratories, with location, proximity, and phone use measured every five minutes, along with a few other sociometric variables.[15] This study covers more than 330,000 hours of human interaction.

http://realitycommons.media.mit.edu

> **Badge Data Set:** One month within a white-collar workplace, with location, communication, and body language measurements taken every sixteen milliseconds, along with precise work flow and task measurements.[16]

Anonymized data, visualizations, code, documentation, and papers can be found at http://realitycommons.media.mit.edu. These data sets were obtained under U.S. federal human subjects law.[17]

These particular living labs give a detailed picture of swaths of American life, but what about life in the developing world, where the vast majority of people live? On May 1, 2013, I hosted the public unveiling of Data for Development, which is perhaps the world's first true big-data commons: It describes mobility and call patterns along with economic, census, political, food, poverty, and infrastructure data for the entire African country of Ivory Coast. These data are now available from http://www.d4d.orange.com/home.

These aggregated anonymous data were donated by the mobile carrier Orange, with help from the University of Louvain (Belgium) and my research group at MIT, and in collaboration with Bouake University (Ivory Coast), the United Nations's Global Pulse, the World Economic Forum, and the GSMA, which is the world trade association of mobile telephone companies. In the final chapter of this book we will see how this data commons is already being used to help revolutionize government and public services in the Ivory Coast.

Plan for the Book

The goal of this book is to explain how social physics brings to-gether big data about human behavior and social science theory to create a practical science that can be—and, in fact, already is being—applied in many real-world settings. In the first part I lay some theoretical groundwork through examples that illustrate the two most important concepts in social physics:

- *Idea flow* within social networks, and how it can be separated into exploration (finding new ideas/strategies) and engagement (getting everyone to coordinate their behavior).
- *Social learning,* which is how new ideas become habits, and how learning can be accelerated and shaped by social pressure.

This part of the book also describes how we can use digital bread crumbs to yield accurate, practical measures of concepts such as social influence, trust, and social pressure. This technique then enables us to measure idea flow within social networks and to deploy incentives that shape the pattern of social learning in real-world situations. I use examples from online social networks, health, finance, politics, and consumer purchasing behavior in order to illustrate the workings of social physics.

In the second part of the book I use various types of real-world examples to demonstrate how social physics has been used to make organizations more flexible, creative, and productive. Examples in-clude research laboratories, creative advertising departments, back-room support operations, and call centers.

The third part examines social physics on a much larger scale, namely for cities. Here my main focus is how we can use social physics to reengineer our cities to be more efficient as well as more creative and productive.

In the final section I discuss how social physics applies to our social institutions. Here I explore the role of government and the structure of law and regulation in a data-driven society and suggest changes to privacy and economic regulation.

Along the way, I hope that the reader will learn the social physics way of thinking. In many respects, this new approach is similar to economics, because of its quantitative, predictive character. Indeed, much of the language I use in this book draws from economics. But rather than study how economic agents work and how economies function, social physics seeks to understand how the flow of ideas turns into behaviors and action. Put another way, social physics is about how human behavior is driven by the exchange of ideas—how people cooperate to discover, select, and learn strategies and coordinate their actions—rather than how markets are driven by the exchange of money.

Social physics also shares some surface resemblance to other academic domains, such as the cognitive sciences. The contrast between most cognitive science and social physics is quite important, however. Rather than focusing on individual thoughts and emotions, social physics focuses on social learning as the major driver of habits and norms. A fundamental assumption is that learning from examples of other people's behavior (and the relevant contextual features) is a major and likely dominant mechanism of behavior change in humans. Because it does not try to capture internal cognitive processes, social physics is inherently probabilistic, with an irreducible kernel of uncertainty caused by avoiding the generative nature of conscious human thought.

Data-Driven Societies: Promethean Fire

The social physics that is emerging brings together branches of economics, sociology, and psychology, along with network, complexity, decision, and ecology sciences and fuses them together using big data. By creating social systems that look beyond aggregates such as markets, classes, and parties, and instead examining the detailed patterns of idea exchanges, I show how we can begin to build a society that is better at avoiding market crashes, ethnic and religious violence, political stalemates, widespread corruption, and dangerous concentrations of power. The first steps are to begin setting scientific, reliable policies for growth and innovation, and to institute information and legal architectures for the protection of privacy and public transparency. Such measures can give us unprecedented instrumentation of how our policies are performing, so we can know when we are being hoodwinked or trod upon, and can take action to quickly and effectively address the situation.

This vision of a data-driven society implicitly assumes that the data will not be abused. However, the ability to see the details of the market, of political revolutions, and to be able to predict and control them is a case of Promethean fire—it could be used for good or for ill. In short, to achieve the exciting possibilities of a data-driven society, we require what I have called the New Deal on Data: workable guarantees that the data needed for public goods are readily available while at the same time protecting the citizenry.[18] Maintaining protection of personal privacy and freedom is critical to the success of any society.

Over the last five years I have co-led a discussion among leading politicians, CEOs of multinational corporations, and public advocacy groups around the world aimed at guaranteeing such individual freedoms. The result is a New Deal on Data that is being developed in the commerce regulations of the United States, the European Union states, and other countries.[19] These changes are beginning to give individuals unprecedented control over data that are about them while at the same time providing for increased transparency and insight in both the public and private spheres.

While these changes will help protect citizens from companies, they do little to protect against the government itself. In June 2013, massive U.S. surveillance of phone records and Internet data was disclosed by former NSA contractor Edward Snowden, who called these activities the "architecture of oppression." We need a renewed public debate on the balance between personal privacy and the government's collection and use of personal data—the New Deal on Data must extend to governments as well. We will also need to adopt computer and communication technologies that make it difficult for governments to overreach themselves.

Another challenge is the need for more controlled experimentation in our social systems. Today governments and companies launch new policies and systems based on very thin evidence. The scientific method as currently practiced in the social sciences is failing us and threatens to collapse in an era of big data.[20] Is coffee good for us or bad? How about sugar? With billions of people consuming these products for over a century, we ought to have the answers. Instead, we have "scientific" opinion that seems to change every day. We need to revive the social sciences by constructing living labs to test and prove ideas for building data-driven societies.

Our society has already begun a great journey that will rival revolutions such as printing and the Internet. For the first time, we will have the data required to really know ourselves and understand how our society evolves. By better understanding ourselves, we can potentially build a world without war or financial crashes, in which infectious disease is quickly detected and stopped, in which energy, water, and other resources are no longer wasted, and in which governments are part of the solution rather than part of the problem. To reach these goals, however, we first need to understand social physics, and then we need to decide as a society what we value most and what we are willing to change to get there.

LANGUAGE

•

Many words have both common English meanings and more technical economics and scientific meanings. To avoid confusion here are some quick definitions:

Engagement: Engagement is social learning, usually within a peer group, that typically leads to the development of behavioral norms and social pressure to enforce those norms. In companies, work groups with a high rate of idea flow among the members of the work group tend to be more productive.

Exploration: Exploration is the process of searching out new, potentially valuable ideas by building and mining diverse social networks. In companies, work groups that have a high rate of idea flow from outside the work group tend to be more innovative.

Idea: An idea is a strategy (an action, outcome, and features that identify when to apply the action) for instrumental behavior. Compatible, valuable ideas become "habits of action" used in "fast thinking" responses.[21]

Idea flow: The propagation of behaviors and beliefs through a social network by means of social learning and social pressure. Idea flow takes into account the social network structure, the strength of the social influence between each pair of people, as well as individual susceptibility to new ideas.

Information: Information is an observation that may be incorporated into a belief or used to build up an idea.

Interaction: Interaction includes both direct (e.g., a conversation) and indirect (e.g., overhearing a conversation) behaviors.

Social influence: Social influence is the likelihood that one person's behavior will affect the behavior of another person.

Social learning: Social learning consists of either: (1) learning new strategies (e.g., context, action, outcome) by observation of other people's behavior, including learning from memorable stories; or (2) learning new beliefs through experience or observation.

Social network incentive: A social network incentive is an incentive to alter the pattern of exchanges between pairs of people.

Social norm: A social norm is a set of compatible strategies that all parties agree produces the best exchange value. Norms are usually developed by social learning and spread by social pressure.

Social pressure: Social pressure is the negotiating leverage one person can exert upon another, which is limited by the exchange value between them.

Society: Social physics assumes that human societies are mostly made up of networks of exchanges between individuals instead of describing society as composed of classes or markets.

Strategy: A strategy is the combination of the features that identify a situation, the potential actions one can take in this situation, and the expected outcomes of those actions.

Trust: Trust is the expectation of continued, stable exchange value.

Value: I will refer to the "value" of an exchange relationship as the extent to which the exchanges satisfy social and personal goals, including, for example, utility, curiosity, and social support.

PART ONE

Social Physics

•

· 2 ·

Exploration

HOW CAN WE FIND GOOD IDEAS
AND MAKE GOOD DECISIONS?

The standard story about innovation and creativity is that there are a very few superbright people who have the almost magical ability to think up great ideas, and the rest of us have occasional lucky breaks. But that is not what I see. Instead, I see that the best ideas come from careful and continuous social exploration.

Here at MIT I occupy a rather unique position in the world, at the crossroads of several broad streams of humanity. The most obvious is that I rub shoulders with many of the best researchers in the world, my colleagues here in the Boston area. Next are the visionary business leaders who come to MIT to speak in my entrepreneurship classes or who sponsor my research. Through the World Economic Forum I have the opportunity to meet and discuss new ideas with political leaders from around the globe. Through the MIT Media Lab I get to engage with many of society's most promising up-and-coming artists. And finally there are the students themselves: humanity's best and brightest from every corner of the globe.

Surprisingly, they are pretty much just normal people. Some of them have skills that they have practiced to the point of world-class expertise. But that is not the source of their new ideas. As Steve Jobs put it:

> Creativity is just connecting things. When you ask creative people how they did something, they feel a little guilty, because they didn't really do it, they just saw something. It seemed obvious to them after a while. That's because they were able to connect experiences they've had and synthesize new things.[1]

The most consistently creative and insightful people are explorers. They spend an enormous amount of time seeking out new people and different ideas, without necessarily trying very hard to find the "best" people or "best" ideas. Instead, they seek out people with *different* views and *different* ideas.

Along with this continuous search for new ideas, these explorers do another interesting thing: They winnow down their most recently discovered ideas to the best ones through their habit of bouncing them off of everyone they meet—and remember that they meet many different sorts of people. Diversity of viewpoint and experience is an important success factor when harvesting innovative ideas. The ideas that provoke reactions of surprise or interest from a wide range of people are the keepers. These are the ideas that are harvested, assembled into a new story about the world, and used to guide actions and decisions.

The most productive people are constantly developing and testing a new story, adding newly discovered ideas to the story and then trying it out on everyone they meet. Like sculpting raw clay into a beautiful statue, over time their story becomes more and more compelling. Finally they decide that it is time to act on it, to bring it into the light and test it against reality. To these people, the practice of harvesting, winnowing, and sculpting with ideas feels like play. In fact, some of them call it "serious play."[2]

The main work of science, art, or leadership is the same: developing a compelling story about the world and then deciding to test it against reality. In science, stories are tested against real-world behavior; in the arts, against their ability to influence the ongoing cultural dialogue; and in management, against their success in business or government.

But how does this exploration process—searching for ideas and then winnowing them down to just a few good ones—result in a harvest of ideas that produces good decisions? Is this just a random recombination of ideas with little contribution from our individual intelligences, or are there strategies that are critical to successful exploration?

Since this exploration process is fundamentally a search of one's social network, a good place to start answering this question is by investigating the role of social interactions in how we find new ideas and use them to make decisions.

Studies of primitive human groups reinforce the idea that social interactions are central to how humans harvest information and make decisions; ethnologists have found that almost all decisions affecting the group as a whole are made in social situations.[3] The major exception to this pattern, in both humans and other animals, is when making decisions extremely rapidly is required, in situations such as battles or emergencies.[4]

A first guess about why humans evolved to use social decision making is that there is an advantage in pooling ideas among many different people. The basic concept is that by pooling ideas, we can get an average "wisdom of the crowd" judgment that will be better than the individual judgments. This sort of idea pooling was popularized some years ago by James Surowiecki's book on "collective intelligence," and it is the basic intuition that motivates secret ballot voting, "likes" and "star" ratings on social media, and download counts on Web pages.[5]

However the evidence is that this idea-pooling approach only works for estimation problems as long as there is no social interaction. In other words, it assumes that all the people in the crowd will act independently. The moment social interaction occurs, however, all bets are off: people begin influencing other people,[6] and that results in panics, bubbles, and fads.[7] Pooling works for simple estimation problems because without social interactions the pieces of information are sufficiently independent that we can use simple mathematics to combine them, i.e., just take the average answer or the median answer.

Unfortunately, there is no easy way to pool more complex strategic information. And yet there is hope. When field biologists observe animal populations, they see that social learning—e.g., copying successful individuals—can improve accuracy in foraging decisions, mate choices, and habitat selections.[8] In humans, the social learning strategy of feeding back the best current idea—that is, a constrained, artificial sort of social interaction that interleaves periods of idea harvesting with periods when experts evaluate the ideas—produces a wisdom of the crowd effect that works even for small groups.[9] In both animals and humans, however, these wisdom of the crowd effects only work as long as the individuals are sufficiently diverse in their decision-making strategies.[10] It seems that the key to harvesting ideas that lead to great decisions is to learn from the successes and failures of others and to make sure that the opportunities for this sort of social learning are sufficiently diverse.

Social Learning

But how in the world does one go about discovering sufficiently diverse ideas? To understand which patterns of social learning generate crowd wisdom, we must first understand the details of how to use social learning to find the best ideas. To illustrate how crowd wisdom can come about, I will walk through some research I did with postdoctoral student Yaniv Altshuler and PhD student Wei Pan on the eToro social network and reported in my *Harvard Business Review* article "Beyond the Echo Chamber."[11]

First, a little background: eToro is an online financial trading platform for day traders, and perhaps its most interesting feature is that it incorporates a social network platform known as OpenBook. In OpenBook, social network users can easily look up other users' trades, portfolios, and past performances, but can not see whom other users copy. Users can place two main types of trades in eToro:

> *Single trade:* To place a normal trade by themselves.

> *Social trade:* To place a trade that exactly copies another user's one single trade, or to follow all of another user's trades automatically.

Most users open up their trading ideas to let other people follow them. They can make quite a bit of money by revealing their

trades in OpenBook, because each time someone decides to copy
another user, that user gets paid a small amount by eToro. Users
typically also select several other users to follow.

During part of 2011 we collected data on euro/dollar trading
from 1.6 million users of eToro. From this data set we were able to
examine almost 10 million financial transactions. The fascinating
and important thing about this example is that we can actually see
the social learning going on, track the effect that this learning has
on people's actions, and measure whether or not their action was
profitable. In short, this network provides us with a god's-eye view
of social learning, i.e., how the detailed exchanges between individual humans affect both their behavior and the final financial
outcome. There are few, if any, other data sets in which we can see
social exploration so clearly and determine which patterns of social
learning work the best.

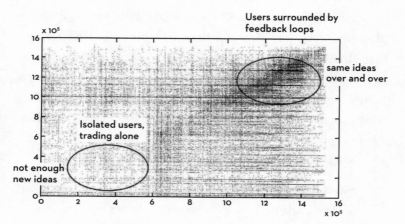

Figure 2. Each dot is a social trade, in which one of the 1.6 million users along the horizontal
axis copied another user, and the vertical axis shows which user they are copying. Two regions
are circled: one where users are relatively isolated and antisocial, and another where users are
all copying each other.

Figure 2 shows the pattern of social learning among the eToro users: Each dot shows which user (horizontal axis) copied which other user (vertical axis). The god's-eye view shown in Figure 2 is something that no individual ever sees: the global pattern of learning. Perhaps the most obvious feature in Figure 2 is that there is still a lot of white space between the dots. This means that most users copy only a few others, and many do not copy anyone at all. What we see here is a social network with sparse connections between individual traders and those they copy. As a result, new trading strategies propagate from user to user along the social network as individual traders decide to copy some other user, and they in turn are copied by other traders.

This figure also makes it clear that there is a lot of variation in the amount of social learning going on. There is one region that is almost covered with dots, meaning that there is a dense network of social learning among these traders. The second region, the one with very few dots, shows that there is little social learning going on among these traders. For most of Figure 2, however, there are a medium number of dots, meaning that most traders fall somewhere between these two extremes of social learning.

What does this mean for individual traders? Some people clearly have impoverished opportunities for social learning because they have too few links to others. Others are embedded in a web of feedback loops, so that they hear only the same ideas over and over again, while most users have a middling number of opportunities for social learning. Each person trading on eToro has a different pattern of exploration, and as a consequence has a different set of ideas to work with.

Idea Flow

What patterns of exploration and social learning produce the best outcomes? We discovered the answer when Yaniv plotted all of the traders' returns on investment against the rates of idea flow among the traders; that is, the rates at which new trading strategies propagate from user to user along the social network as individual traders decide to copy some other user, and they in turn are copied by other traders.[12]

Figure 3 shows how return on investment varies with rate of idea flow within the eToro social network. In this figure, each dot is the average performance of all eToro social traders averaged over an entire day. Altogether, almost 10 million trades were used to calculate the data shown in this graph. The rate of idea flow is shown on the horizontal axis, while the vertical axis compares the

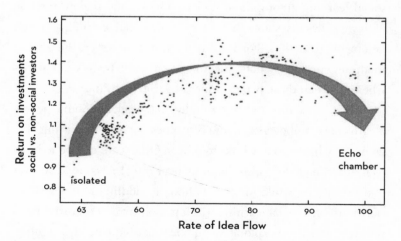

Figure 3. Each dot is the average performance of all eToro social traders averaged over a day. The vertical axis compares the return on investment of social trading (corrected to remove changes in the market) and the horizontal axis shows the rate of idea flow within the eToro social network. With the proper rate of idea flow, return on investment increases 30 percent over individual trading.

returns on investment, corrected to remove the effects of overall ups and downs in the market.[13]

A quick glance at the rates of idea flow within the eToro social network reveals a large range, from isolated individual traders at one end to traders trapped in an echo chamber at the other end. Looking at the profitability of the isolated echo chamber, and in-between groups, it is clear to see that the effect of social learning is enormous. When traders had the right balance and diversity of ideas in their social network, their return on investment increases 30 percent over individual traders.[14]

For this digital trading environment, the wisdom of the crowd resides in between the extremes of isolation and the herd behavior seen when the social network is an echo chamber. This intermediate zone is where social learning, that is, copying successful people, yields real rewards. In later chapters we will see this wisdom of the crowd idea returning in companies, cities, and social institutions.

This wisdom effect is not unexpected; there are hints of this in ape troops[15] and small human groups;[16] it is also seen in simulations of networks of computer learning algorithms[17] and in mathematical models of social learning.[18] What postdoctoral student Erez Shmueli, Yaniv, and I found is that a community of social learners spontaneously forms what is called a scale-free[19] fractal network—one whose connections are systematically more diverse than merely random—and, in addition, that connections in the network change over time in this same scale-free fractal manner.[20] As the pattern of connections between learners becomes optimal, the performance of the entire crowd improves dramatically. The result is a fractal dance of learning that spins ideas into wisdom.

So what exactly—in English—is idea flow? The spread of an idea through a social network is similar to the spread of the flu. In the case of the flu virus, with each exchange between an infected person and a new person, there is a certain chance that the flu virus will infect the new person. If there is a lot of interaction and the new person is susceptible to the virus, then he will likely get the flu. If most people are susceptible, then the virus will eventually spread through most of the population.

The flow of ideas is similar. The process of social learning implies that if there is a lot of interaction between someone showing the behavior (the role model) and a new person, and if the new person is susceptible, then it is likely that this new idea will take root and change the new person's behavior. Susceptibility depends on several factors, including whether the role model is sufficiently similar to the new person that the new behavior is likely to be useful; that the level of trust between the role model and the new person is high; and the consistency between the new idea and previously learned behaviors. As a result, the flow of ideas can sometimes be quite slow, rather different from the viral marketing that advertising agencies like to talk about.[21]

So our measure of idea flow is the chance that a person's behavior will change when a new idea has appeared in their extended social network. It's a lot like the chance of getting the flu during flu season, only ideas usually don't spread as far or as fast as the flu. In fact, the only way we have reliably been able to trigger rapid cascades of idea flow is through the use of social network incentives, as we will see in the next few chapters.

Idea Flow and Decision Making

This eToro example makes it very clear that the rate of idea flow is a critical measure of how well the social network functions in collecting and refining decision strategies. In later chapters we will see that the rate of idea flow can also be used to predict productivity and creative output.

But what can a single individual do to increase rate of idea flow in their part of their social network? Fortunately, there are many ways. In 1985, Bob Kelly of Carnegie Mellon University launched the now famous Bell Stars study.[22] Bell Laboratories, a premier research laboratory, wanted to know more about what separates a star performer from the average performer. Is it something innate or can star performance be learned? Bell Labs already hired the best and the brightest from the world's most prestigious universities, but only a few lived up to their apparent potential for brilliance. Instead, most hires developed into solid performers but did not contribute substantially to AT&T's competitive advantage in the marketplace.

What Kelly found was that star producers engage in "preparatory exploration"; that is, they develop dependable two-way streets to experts ahead of time, setting up a relationship that will later help the star producer complete critical tasks. Moreover, the stars' networks differed from typical workers' networks in two important respects. First, they maintained stronger engagement with the people in their networks, so that these people responded more quickly and helpfully. As a result, the stars rarely spent time spinning their wheels or going down blind alleys.

Second, star performers' networks were also more diverse. Average performers saw the world only from the viewpoint of their job, and kept pushing the same points. Stars, on the other hand, had people in their networks with a more diverse set of work roles, so they could adopt the perspectives of customers, competitors, and managers. Because they could see the situation from a variety of viewpoints, they could develop better solutions to problems.

There are also ways people can change their personal habits in order to increase idea flow. In 2004, PhD student Tanzeem Choudhury and I used sociometer badges to track interactions among four research groups for a period of two weeks, producing a millisecond-by-millisecond record for an average of sixty-six hours of data per subject (details about these badges can be found in the Reality Mining appendix).[23]

What we found was that individuals who adopted an energetic, engaging interaction style that created more interactive conversations ended up being more important to idea flow in the social network.[24] This is just what I see when I look at the most productive people in the world: They are continually engaging with others in order to harvest new ideas, and this exploratory behavior creates better idea flow.

Idea flow also depends on the mix of social learning and individual learning. For example, when people see others adopting strategies similar to their own, they often become more confident, and they are then likely to increase their investment in that particular strategy. People's decisions are a blend of personal information and social information, and when the personal information is weak, they will tend to rely more on social information. In a situation where people are uncertain, the confidence-enhancing effect of social learning becomes larger.[25] This makes perfect sense: When people don't know what is going on, they can learn by spending more time looking at what others are doing.

Unfortunately, this can also lead to overconfidence and group-think, because the mechanism of social learning only improves decision making when people have different individual information. So in situations when the outside information sources (e.g., magazines, TV, radio) become too similar, then groupthink becomes a real danger.

Similarly, when there are feedback loops in the social network, then the same ideas circle around and come back again and again. But because ideas usually change slightly as they go from person to person, they may not be recognized as repetitions of the same ideas. It is easy to believe that everyone has independently arrived at similar strategies, and again become more confident than is warranted. This echo chamber overconfidence effect is a source of fads and financial bubbles.

Sometimes these echo chamber situations do not end pleasantly, as is often the case with financial bubbles or panics. The region of dense feedback loops in Figure 2 is, in fact, a sort of financial bubble. It turned out that a particular trader in Latvia had a long winning streak, and over time, people began to copy him, and then people began to copy the copiers, and on and on. Quietly but quickly, social learning produced a large "organization" that was exploiting the Latvian strategy.

The fact that the traders couldn't see the whole network of social connections, however, meant that they didn't know that they were all copying one trader in Latvia. They thought they were following several different "gurus" who had somehow independently come up with similar strategies. Because there seemed to be so much independent support for the strategy, these traders became far too confident in it. Unfortunately, all gurus lose eventually, and for the people whose portfolios consisted entirely of reflections of this one Latvian trader, the result was a disaster. Pop goes the bubble.

Tuning Networks

Because idea flow takes into account the variables of a social network structure, the strength of social influence between people, and individual susceptibilities to new ideas, it also serves another vital role: It gives a reliable prediction of how changing any of these variables will change the performance of all the people in the network. Thus, this mathematically derived concept of idea flow allows us to "tune" social networks in order to make better decisions and achieve better results.

For example, what can be done when the flow of ideas becomes either too sparse and slow or too dense and fast? Within the eToro digital finance world, we have found that we can shape the flows of ideas between people by providing small incentives or nudges to individuals, thus causing isolated traders to engage more with others and those who were too interconnected to engage less and explore outside their current contacts.

In one experiment with the eToro investors, Yaniv Altshuler and I were able to use this method to tune the social network so that it remained in the healthy wisdom of the crowd region, where traders had sufficiently diverse opportunities for social learning but were not caught up in echo chambers, where loops of social learning cause the same ideas to recirculate endlessly. As a result of this tuning we were able to increase the profitability of *all* the social traders by more than 6 percent, thus doubling their profitability.[26]

In this example, our tuning worked to break up the echo chamber, reducing the recirculation of currently popular strategies and giving new strategies a chance to catch on. By reducing the rate of idea flow to allow greater diversity, we moved the social network back into its sweet spot and raised average performance. Through managing idea flows, our tuning of the network turned average traders—often the losers in our current financial system— into winners. Good idea flow is money not only in financial networks but (as we will see in subsequent chapters) also in companies and cities.

This tuning concept is applicable to a wide range of networks, not just the eToro one. We have seen the same network structures in news reporters' sources (so we can tell if they are getting all sides of a story), financial controls (to ensure that all sources of fraud have been considered), and ad campaigns (making sure that a sufficiently diverse set of customer opinions has been sampled). As a result, Yaniv and I have created a spin-off company called Athena Wisdom that is now tuning financial and decision-making networks around the world.

Exploration

The eToro and Bell Stars examples give a good picture of how networks of relationships determine the quality of real-world decision making. From here onward I will use the term "exploration" to refer to the use of social networks in harvesting ideas and information. Exploration is the part of idea flow that brings new ideas into a work group or community. There are three main things to remember about exploration:

> *Social learning is critical*: Copying other people's successes, when combined with individual learning, is dramatically better than individual learning alone. When your individual in-

formation is unclear, rely more on social learning; when your individual information is strong, rely less on social learning. These examples also teach important details about good decision making. The power of social learning can be seen in social networks. Increasing your reach and your network's diversity makes it more likely that you can find the best strategies.

Diversity is important: When everyone is going in the same direction, then it is a good bet that there isn't enough diversity in your information and idea sources, and you should explore further.[27] A big danger of social learning is groupthink. How can you avoid groupthink and echo chambers? You have to compare what the social learning suggests with what isolated individuals (who have only external information sources) are doing. If the so-called common sense from social learning is just an overconfident version of what isolated people think, then you are likely in a groupthink or echo chamber situation. In this case, a surprisingly good strategy is to bet against the common sense. In fact, for eToro users this strategy had returns that were reliably better than all but the best human traders.

Contrarians are important: When people are behaving independently of their social learning, it is likely that they have independent information and that they believe in that information enough to fight the effects of social influence. Find as many of these "wise guys" as possible and learn from them. Such contrarians sometimes have the best ideas, but sometimes they are just oddballs. How can you know which is which? If you can find many such independent thinkers and discover that there is a consensus among a large subset of them, then a really, really good trading strategy is to follow the contrarian consensus. For instance, in the eToro network the consensus of these independent strategies is reliably more than twice as good as the best human trader.

In summary, people act like idea-processing machines combining individual thinking and social learning from the experiences of others. Success depends greatly on the quality of your exploration and that, in turn, relies on the diversity and independence of your information and idea sources. The following chapters will show this same process of exploration playing a key role in the creative output of organizations, cities, and society as a whole.

One disturbing implication of these findings is that our hyperconnected world may be moving toward a state in which there is too much idea flow. In a world of echo chambers, fads and panics are the norm, and it is much harder to make good decisions. This suggests that we need to pay much more attention to where our ideas are coming from, and we should actively discount common opinions and keep track of the contrarian ideas. We can build software tools to help us do this automatically, but to do so we have to keep track of the provenance of ideas. Older systems such as copyrights were a first step toward keeping track of idea flow, but much more uniform and lightweight mechanisms are required. I will return to this topic in the final two chapters, where I write about how we can build trust-network systems that protect personal privacy and yet allow for sufficient flow of ideas.

Finally, while I have given a written description here of our research work, at its heart lies sophisticated mathematics. Yaniv Altshuler, Wei Pan, Wen Dong, and I have created detailed equations that quantify the process of social learning and exploration; these analyses can be used to discover how best to harvest ideas from social networks and make better decisions. Utilizing these equations, we can reliably predict what individuals will choose to do and how good their outcomes will be in situations ranging from companies (Part II of this book), to cities (Part III), to entire countries (Part IV).

These equations are a core part of social physics, and those interested in the equations should take a look at the special topic box on the Mathematics of Social Influence (page 80) as well as the Math appendix.

· 3 ·

Idea Flow

THE BUILDING BLOCKS OF COLLECTIVE INTELLIGENCE

Why do some companies feel energetic and creative while others feel stale and stagnant? Or what about businesses where everyone seems to be working frantically, but there is no sense of cohesion or direction? The standard explanations usually come down to describing the work at some companies as fun or exciting rather than dull and boring, or that some companies are just well managed while others are not.

But this isn't what I see when I look around. Instead, I see that companies have different types of idea flow—and therefore different abilities to learn both from inside and outside their communities. In each case, I think that the causes of excitement, boredom, or craziness have more to do with how tightly coupled people are to each other or how deep the splits are between divisions than with specific management techniques or the nature of the company's work. In others words, it is the rates of idea flow—or the barriers to idea flow—that we must understand if we are to work well together.

I think of organizations as a group of people sailing in a stream of ideas. Sometimes they are sailing in swift, clear streams where the ideas are abundant, but sometimes they are in stagnant pools or terrifying whirlpools. At other times, one person's idea stream forks off, splitting them apart from other people and taking them in a new direction. To me, this is the real story of community and culture. The rest is just surface appearance and illusion.

Idea flow is the spreading of ideas, whether by example or story, through a social network—be it a company, a family, or a city. This flow of ideas is key to the development of traditions, and ultimately of culture. It facilitates the transfer of habits and customs from person to person and from generation to generation. Further, being part of this flow of ideas allows people to learn new behaviors, without the dangers or risks of individual experimentation, and to acquire large integrated patterns of behavior, without having to form them gradually by laborious experimentation.[1]

Each cohesive community has its own stream of idea flow that allows the members to incorporate innovations from other people within it, and even to create a separate culture. Examples of such "communities of practice" are the artisan guilds that flourished in the Middle Ages, today's professional associations, or even the eToro community described in the last chapter.

As we saw in Chapter 2, the right sort of idea flow leads to all of the group members making better decisions than they could have made if they were operating individually. As a consequence of these shared habits, human communities can develop a sort of collective intelligence that is greater than the members' individual intelligence. Engagement with and learning from others, along with the mutual sharing and vetting of ideas, generate the collective intelligence.

Idea flow depends upon social learning, and indeed, this is why social physics works: Our behavior can be predicted from our exposure to the example behaviors of other people. In fact, humans rely so much on our ability to learn from the ideas that surround us that some psychologists refer to us as *Homo imitans*.[2] Through social learning we develop a shared set of habits for how to act and respond in many different situations. The mundane details of our daily lives have their basis in habits; in aggregate, these habits define our society. We drive on the left (or right), wake at 8:00 A.M. (or 6:00 A.M.), and eat with a fork (or chopsticks).

Such social learning is not unique to humans. Other primates, such as chimpanzees and orangutans, also exhibit behavioral cultures in the wild. For example, innovations in food gathering sometimes spread throughout a troop, and this idea flow replaces old habits with new, more efficient methods of harvesting. Even though such idea flows bring innovation into their repertoires of habits, these primate cultures remain simple and static.

One reason human culture grows while ape culture remains stagnant may be that, unlike apes, we occasionally choose to row against the flow of ideas that surround us and dip into another stream. In Chapter 2, we saw that new and more successful behaviors can enter the community when we use our social networks to explore and test new ideas. By harvesting from the parts of our social network that touch other streams, that is, by crossing what sociologist Ron Burt called the "structural holes" within the fabric of society, we can create innovation. When we choose to dip into a different stream, we bring up new habits and beliefs; in some cases, they will help us make better decisions, and our community will thrive.[3] I believe that we can we think of each stream of ideas as a swarm or collective intelligence, flowing through time, with all the humans in it learning from each other's experiences in order to jointly discover the patterns of preferences and habits of action that best suit the surrounding physical and social environment.

This is counter to the way most modern Westerners understand themselves, which is as rational individuals, people who know what they want and who decide for themselves what actions to take in order to accomplish their goals. Could it be that our preferences and methods of action, the very things that define rationality, come from our community as much as from within ourselves? Are we, by the definition of economists, as much collectively rational as we are individually rational?

Habits, Preferences, and Curiosity

To answer this question we need to understand more about how the flow of ideas works, that is, how the example behaviors that surround us end up as our own habits, preferences, and interests. To investigate this question I launched two big data studies, one called Social Evolution and the other, Friends and Family, which contain almost 2 million hours of interaction data covering everyone within two communities for a total of over two years.[4] (For more details, see the Reality Mining appendix, and for papers, data, and visualizations see http://realitycommons.media.mit.edu.)

Habits: Are our habits the result of our personal choices or do they come from the flow of ideas that surrounds us? We know that obesity, smoking, and other health-related behaviors are affected by social learning, and social support is known to be a key factor in an individual's health and well-being. For example, longitudinal studies of participants in the Framingham Heart Study have suggested that social interaction is important in the spreading of behaviors ranging from obesity to happiness.[5] These studies are limited in their ability to help us understand how we acquire health habits, however, because they mostly included only friends and family members, and because many of the data are sparse and retrospective, i.e., infrequent records of people's memories of events and not the real-time, quantitative measurements that are typical of big data.

So to answer the question of how habits form, my research group studied the spread of health behaviors in a tightly knit undergraduate dorm for one year. In the Social Evolution Study, led by PhD student Anmol Madan and myself, with Professor David Lazer helping with design of the experiment and data analysis, we gave all the participating students smartphones with special software so that we could track their social interactions with both close friends and acquaintances. In total, this study produced more than five hundred thousand hours of data and included face-to-face interactions, phone calls, and texting, as well as extensive surveys and weight measurements.[6] These hundreds of gigabytes of data allowed us to examine what goes into the creation of habits.

One particular health behavior that we focused on was weight change and on whether this was more influenced by the behavior of friends or by peers in the surrounding community. For most people usually only a few peers are also friends and instead most are only acquaintances with whom we have relatively little interaction. Since peers and friends are only partially overlapping groups, the results for the two groups can be quite different.

What we found was that weight change showed a very strong association with exposure to peers who gained weight but not to those who lost weight. And further, social interaction with close friends who experienced a weight change showed no significant effect at all. A similar effect was also found when we examined eating habits, with exposure to peers being the key variable.

In this case it wasn't just direct interactions that mattered; it was the amount of all exposure to the behavior of people who gained weight, including both direct interaction and indirect observation. In other words, exposure to overheard comments or casual observation of other people's behavior can drive idea flow just as well as, and in some cases better than, more direct interactions such as conversations, telephone calls, and social media. Idea flow sometimes depends more on seeing what people actually do than on hearing what they say they do.

In fact, exposure to the behavior examples that surrounded each individual dominated everything else we examined in this study. It was more important than personal factors, such as weight gain by friends, gender, age, or stress/happiness, and even more than all these other factors combined. Put another way, the effect of exposure to the surrounding set of behavior examples was about as powerful as the effect of IQ on standardized test scores.

It might be asked how we can know that exposure to the surrounding behaviors actually caused the idea flow; perhaps it is merely a correlation. The answer is in this experiment we could make quantitative, time-synchronized predictions, which make other noncausal explanations fairly implausible. Perhaps even more persuasively, we have also been able to use the connection between exposure and behavior to predict outcomes in several different situations, and even to manipulate exposure in order to cause behavior changes.[7] Finally, there also have been careful quantitative laboratory experiments that show similar effects and in which the causality is certain.[8]

Therefore, people seem to pick up at least some habits from exposure to those of peers (and not just friends). When everyone else takes that second slice of pizza, we probably will also. The fact that exposure turned out to be more important for driving idea flow than all the other factors combined highlights the overarching importance of automatic social learning in shaping our lives.

Preferences: Perhaps eating too much might be something that we would expect to be naturally "absorbed" from the surrounding examples of our peers' behavior: When in Rome we do as the Romans do. But how do the example behaviors that surround us impact beliefs and values that are more considered and rational?

In particular, we were interested in political preferences. How do we choose who to vote for? Do our preferences also come from exposure to those around us? We tackled this question in the Social Evolution experiment by analyzing students' political views during the 2008 presidential election.[9] The question we asked was: Do political views reflect the behaviors that people are exposed to or are they formed more by individual reasoning? By giving these students specially equipped smartphones, we monitored their patterns of social interaction by tracking who spent time with whom, who called whom, who spent time at the same places, and so forth.

We also asked the students a wide range of questions about their interest in politics, involvement in politics, political leanings, and finally (after the election), we inquired which candidate had received their vote. In total, this produced more than five hundred thousand hours of automatically generated data about their interaction patterns, which we then combined with survey data about their beliefs, attitudes, personality, and more.

When sifting through these hundreds of gigabytes of data, we found that the amount of exposure to people possessing similar opinions accurately predicted both the students' level of interest in the presidential race and their liberal-conservative balance. This collective opinion effect was very clear: More exposure to similar views made the students more extreme in their own views.

Most important, though, this meant that the amount of exposure to people with similar views also predicted the students' eventual voting behavior. For first-year students, the size of this social exposure effect was similar to the weight gain ones I described in the previous section, while for older students, who presumably had more fixed attitudes, the size of the effect was less but still quite significant.

But what did not predict their voting behavior? The views of the people they talked politics with, and the views of their friends. Just as with weight gain, it was the behavior of the surrounding peer group—the set of behavior examples that they were immersed in— that was the most powerful force in driving idea flow and shaping opinion. Again, it is important to notice that it wasn't just the number of direct interactions that mattered, but rather, it was the amount of exposure to other people's statements and attitudes, both direct interactions through conversations and indirect interactions through incidental observation. Overheard comments and the observation of other people's behavior are effective drivers of idea flow.

In this case, the picture is more complex, because when politics became a more important topic of discussion, such as just before a televised presidential debate, the students changed who they spent time around. If they were conservative in their political views, they withdrew from places frequented by their more liberal acquaintances. Similarly, if they were leaning liberal, then they avoided places with many conservatives.[10]

It was reassuring to see that there was at least some role for individual preferences, as it seems likely that the crowd they chose to spend time around was selected by how comfortable they were with the crowd's offhand comments and opinions. This process of selective exposure then reinforced their political views. But once they chose a side, their increased exposure to similar ideas continued to shape their thinking so that they eventually became true believers. As Nobel Laureate Daniel Kahneman might have put it, we can consciously reason about which flow of ideas we want to swim in, but then exposure to those ideas will work to shape our habits and beliefs subconsciously.

New Ideas and Information: In the examples of eating habits and political preferences, we saw that exposure, both direct and indirect, was the major factor in forming both habits and preferences. For political views, spending more time with a crowd that felt comfortable shifted exposure to a different flow of ideas in a way that hardened beliefs and habits.

But what about our search for new ideas and information? Do our curiosity and our interests stem from our own individual choices or from the crowd surrounding us? If the latter, then not only the process of selecting and adopting new behaviors, but also the very source of idea flow depends on a community consensus. To investigate the process behind idea discovery, PhD students Wei Pan and Nadav Aharony and I used the Friends and Family study to monitor the adoption of smartphone apps in a community of young families.[11] We gave smartphones to all the adults in this community and put special software on the phones to record who called, e-mailed, or texted whom, who were active friends on social media, who spent time face-to-face, and also where they spent time.

To assess app downloading behavior, we monitored which apps they downloaded on the smartphones, giving us a view of which tools, games, and information sources they were choosing. All combined, these smartphones generated over 1.5 million hours of automatically recorded data, including the participants' app downloads and interpersonal interaction patterns. In addition, we collected hundreds of surveys about their beliefs, attitudes, personality, and other characteristics.[12]

The data from app downloading allowed us to examine the decision-making environment surrounding these choices. We could examine the data to see if they were independent decisions, decisions driven by advertising, or decisions driven by exposure, i.e., interactions with other people who had already downloaded the app.

When we analyzed the hundreds of gigabytes of data from this experiment, we first verified the standard sociology result: People with similar characteristics (e.g., age, gender, religion, employment, etc.) tended to download similar apps. This similarity effect, however, gave us only about 12 percent accuracy in prediction. In contrast, when we analyzed participants' exposure through all channels of communication—e.g., face-to-face interactions (including conversations and incidental observations), phone calls, social media, etc.—we could predict which app a person would download with four times more accuracy. So even in a case that seemed squarely in the domain of conscious decision making, the

predictive power of exposure to surrounding peer behavior dominated. The search for new ideas and information, like the formation of new habits, appears driven primarily by social exposure.

The same type of process seems to guide the process of discovery in online situations as well. Consider this experiment, reported
in the highly regarded Public Library of Science journal *PLOS
ONE*,[13] in which we examined data from an online cultural market
in which fourteen thousand users interacted to download digital
music.[14] Typical of online sites, more popular songs were listed first
and the number of downloads for each song was also listed (although there were several different Web site configurations tested).
Just as with the app adoption example discussed earlier, we found
that users' behavior could be quite accurately explained by a simple statistical model of social influence. The decision to sample
the music was indeed driven by online social influence, such as the
song's ranking and the number of downloads.

The app and music cases are different from the health habits or
political preferences, however. In both cases we could make good
predictions about what people sampled but not what they would
actually use or purchase. The effect of social exposure was infor-
mational, guiding their search for new apps or new music, but not
normative. In other words, the particular app or music sampled
often did not turn into a habit.

The bottom line: In these three examples—health habits, political preferences, and consumer consumption—exposure to the
behavior of peers, both direct and indirect, predicted idea flow.
The effects of exposure to peer behaviors are roughly the same size
as the influence of genes on behavior or IQ on academic perfor-

mance. Moreover, in each case it appears that exposure to surrounding peer behaviors is the largest single factor driving idea flow.

Perhaps this is because learning from surrounding example behaviors is much more efficient than learning solely from our own experiences. Mathematical models of learning in complex environments suggest that the best strategy for learning is to spend 90 percent of our efforts on exploration, i.e., finding and copying others who appear to be doing well.[15] The remaining 10 percent should be spent on individual experimentation and thinking things through.[16]

The logic behind this is simple: If somebody else has invested the effort to learn some useful behavior, then it is easier to copy them than to think it through all over again. As a simple example: If we have to use a new computer system, why read the manual if we can watch someone else who has already learned to use the system? People overwhelmingly rely on social learning and are more efficient because of it.

It is significant that we found that people choose to change their environment in order to change the behaviors they were exposed to. One way to interpret all of these findings is that the more people want to learn from a particular peer group, i.e., the more they want to fit in, then the more time they spend around them.

Harnessing the power of exposure to drive idea flow can be used to cause desirable behavior changes. For instance, a similar exposure-based mechanism is likely to be a big part of the reason behind the success of team-based weight-loss programs such as Weight Watchers and participatory TV shows such as *The Biggest Loser*, on which overweight contestants battle to lose the most weight.

As the work of Stanley Milgram on social conformity demonstrated, when our peers are all doing the same thing, be it gaining or losing weight, or even doing something outrageous such as doling out electric shocks, the uniformity of the example behaviors around us strongly influences both unconscious habits and conscious decisions. Many commentators have observed that the power of social influence can lead people to both good and bad behaviors, and influence our behaviors to an extent that is scarcely believable. In the next chapter, we will see how we can use social network incentives to change exposure, and so harness this power to shape the flow of ideas. Indeed, we will see that providing social network incentives to change idea flow is a far more powerful method of changing behaviors than the traditional method of using individual incentives.

Habits versus Beliefs

The Social Evolution and Friends and Family studies paint a picture of humans as sailors. We all sail in a stream of ideas, ideas that are the examples and stories of the peers who surround us; exposure to this stream shapes our habits and beliefs. We can resist the flow if we try, and even choose to row to another stream, but most of our behavior is shaped by the ideas we are exposed to. The idea flow within these streams binds us together into a sort of collective intelligence, one comprised of the shared learning of our peers.

This is an uncomfortable picture for most of us, however. Where are our principles? Our morals? Where, even, is our reasoning and system of beliefs? In order to understand the role of reason in idea flow, we need to untangle and analyze the complicated question of how habits and beliefs are created.

A key part of this puzzle can be found in the work of psycholo-

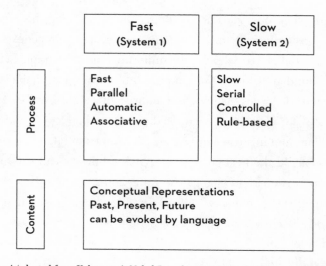

Figure 4 (adapted from Kahneman's Nobel Prize lecture): Humans have two ways of think-ing: an older capability based on association and experience ("fast") and a newer capability based on attentive, rule-based thinking ("slow").

gist Daniel Kahneman and artificial intelligence pioneer Herb Simon, both Nobel Prize winners.[17] As illustrated in Figure 4, each embraced a model of a human mind with two ways of thinking: a fast, automatic, and largely unconscious mode, along with a slow, reasoning, and largely conscious one.[18] A thumbnail sketch of these is that fast thinking drives our habits and intuitions, largely by using associations among ideas learned from our experiences and those we've learned by observing others. In contrast, the slow mode uses reasoning, combining our beliefs in order to reach new conclu-sions. (For more detail, see the Fast, Slow, and Free Will appendix.)

It surprises most people to learn that fast thinking is better than slow thinking for many tasks.[19] Whenever a problem is complex and involves trade-offs between different goals, the association mechanisms used in fast thinking usually outperform the slower reasoning mechanisms. This is especially true when there is limited time to make a decision. For this reason, many scientists believe that the vast majority of our daily behavior is due to fast thinking—we literally don't have the time to think things through using slow thinking.

Interestingly, fast thinking also seems to play an important role in creating healthy societies. Psychological studies have shown that the snap judgments of people are more altruistic and cooperative than the decisions made slowly and thoughtfully.[20] Examples such as the reactions of spectators at the Boston Marathon bombings, or of neighbors after the recent Oklahoma tornadoes show that this fast-thinking core of human nature plays an important role in building strong communities.

While we may make a high-level, conscious decision to engage in some activity, many of the activities themselves are highly practiced and automatic, driven by fast thinking and largely out of the spotlight of our attention. The largely automatic nature of our lives is most visible when we are expert at an activity, such as performing the daily routines of life, engaging in social chitchat, or physical activities, such as driving or riding a bike. We are typically hard-pressed to explain exactly what we did or why we did these habitual activities, because we were simply on autopilot.

But what do fast and slow thinking have to do with streams of ideas and collective rationality? The answer is that the process of learning is different for these two ways of thinking, and they are also used differently in daily life; these differences matter when trying to understand how communities build a collective intelligence.

For factual beliefs (e.g., "The dinner starts at 7:00 P.M."), a single exposure from a trusted peer is usually sufficient to convert a person to that belief. In contrast, to change habitual behaviors, preferences, and interests, it usually requires several exposures within a short period of time. For instance, if everyone in a work group starts drinking green tea rather than coffee, the odds are good that others will pick up the green tea habit as well. Multiple exposures showing that a new behavior has a good outcome (such as social approval) are needed before we are likely to pick up the habit as our own.

My experiments suggest that the continual exploratory behavior of humans is a quick learning process that is guided by apparent popularity among peers. In contrast, adoption of habits and preferences is a slow process that requires repeated exposure and perceptual validation within a community of peers. Our social world consists of the rush and excitement of new ideas harvested through exploration, and then the quieter and slower process of engaging with peers in order to winnow through those ideas, to determine which should be converted into personal habits and social norms.

Over time we develop a shared set of habits for how to act and respond in many different situations, and these largely automatic habits of action account for the vast majority of our daily behavior. As Nobel Laureate Herb Simon put it, our rational, conscious thinking is the program that invokes the habits of action that take care of all the details of daily life, just as computer programs have subroutines that handle frequently used computations.

Collectively Rational, Not Individually Rational

Near the end of the 1700s, philosophers began to declare that humans were rational individuals. People were flattered by being recognized as individuals, and by being called rational, and the idea soon wormed its way into the belief systems of nearly everyone in upper-class Western society. Despite resistance from Church and State, this idea of rational individuality replaced the assumption that truth only came from god and king. Over time, the ideas of rationality and individualism changed the entire belief system of Western intellectual society, and today it is doing the same to the belief systems of other cultures.

As we have seen so far in this book, new data are changing this argument, and we are now coming to realize that human behavior is determined as much by social context as by rational thinking or individual desires. Rationality, as the term is used by economists, means that we know what we want and act to get it. But I think that my research shows that both people's desires and their decisions about how to act are often, and perhaps typically, dominated by social network effects.

Recently, economists have moved toward the idea of bounded rationality, which means that we have biases and cognitive limitations that prevent us from realizing full rationality. Our dependence on social interactions, however, is not simply a bias or a cognitive limitation. As we saw in Chapter 2, social learning is an important method of enhancing individual decision making. Similarly, we will see in the next chapter that social influence is central to constructing the social norms that enable cooperative behavior. Our ability to survive and prosper is due to social learning and social influence at least as much as it is due to individual rationality.

These data tell us that what we want and value, as well as how we choose to act in order to obtain our desires, are a constantly evolving property of interactions with other people. Our desires and preferences are mostly based on what our peer community agrees is valuable rather than on rational reflection based directly on our individual biological drives or inborn morals.[21] For instance, after the Great Recession of 2008, when many houses were suddenly worth less than their mortgages, researchers found that it only took a few people walking away from their houses and mortgages to convince many of their neighbors to do the same thing.[22] A behavior that had previously been thought nearly criminal or immoral, i.e., purposely defaulting on a mortgage, now became common. Using the terminology of economics, in most things we are collectively rational, and only in some areas are we individually rational.

Common Sense

The collective intelligence of a community comes from idea flow; we learn from the ideas that surround us, and others learn from us. Over time, a community with members who actively engage with each other creates a group with shared, integrated habits and beliefs. When the flow of ideas incorporates a constant stream of outside ideas as well, then the individuals in the community make better decisions than they could on their own.

This idea of a collective intelligence that develops within communities is an old one; indeed, it is embedded in the English language. Consider the word "kith," familiar to modern English speakers from the phrase "kith and kin." Derived from old English and old German words for knowledge, kith refers to a more or less cohesive group with common beliefs and customs. These are also the roots for "couth," which means possessing a high degree of sophistication, as well as its more familiar counterpart, "uncouth." Thus, our kith is the circle of peers (not just friends) from whom we learn the "correct" habits of action.

Our ancestors understood that our culture and the habits of our society are social contracts, and that both depend primarily upon social learning. As a result, most of our public beliefs and habits are learned by observing the attitudes, actions, and outcomes of peers, rather than by logic or argument. Learning and reinforcing this social contract is what enables a group of people to coordinate their actions effectively.

So even though today's society tends to glorify the individual, the vast majority of our decisions are shaped by common sense, the habits and beliefs we have in common with our peers, and these common habits are shaped by interactions with other people. We learn common sense almost automatically, by observing and then copying the common behaviors of our peers. It is through these collective preference and decision mechanisms that we come to automatically behave politely at parties, respectfully at work, and passively in public transit.[23] It is the idea flow within a community that builds the intelligence that makes it successful.

Engagement

HOW CAN WE ALL
WORK TOGETHER?

In the last two chapters I've explored where ideas come from and how they turn into behaviors, including shared habits. The ability to work together, though, goes beyond simple idea flow within a community; it also includes striking a bargain between individuals to adopt behaviors that are synchronized and compatible. Working together also requires more than shared habits; it requires habits that result in cooperation. We still have to understand how we can get people to work together. How do we come to adopt habits of action that mesh together like pieces in a puzzle, allowing many people to work toward the same goal?

The ability to work as a group is older than humanity. For instance, mountain gorillas decide when to end an afternoon siesta by using "close call" signals.[1] When everyone in the group has been heard from, and the signaling reaches a certain intensity, then the rest period is over. Likewise, capuchin monkeys use trilling sounds to cooperatively decide when and where the troop should move.[2] Monkeys at the leading edge of the troop trill the most, encouraging others to follow the path they have found, and others take up the trilling in order to coordinate everyone's movements.

Similar patterns of social decision making are common in many animals and virtually all primates. The signaling mechanisms vary from vocalization to body posture to head movements, but the structure of the decision-making process remains pretty much the same: cycles of signaling and recruitment, until a tipping point is reached when everyone in the group accepts that a consensus has been reached.[3] Some evolutionary theorists think that this type of "social voting" process could be the most common type of decision making in social animals, in part because it is very good at accounting for the cost-benefit trade-offs of everyone in the group. In addition, this type of consensus process typically avoids extreme decisions, making it more likely that the entire group will follow.

These same types of patterns are found in human organizations. In Bob Kelly's Bell Stars study, in which he looked at the difference between average and star performers within Bell Laboratories, researchers found that the star performers encouraged their work groups to behave in exactly this sort of social voting manner.[4] Average performers thought teamwork meant doing their part on the team. Star performers, however, saw things differently: They pushed everyone on the team toward joint ownership of goal setting, group commitments, work activities, schedules, and group accomplishments. That is, star performers promoted synchronized, uniform idea flow within the team by making everyone feel a part of it, and tried to reach a sufficient consensus so that everyone would willingly go along with new ideas.

Synchronization and uniformity of idea flow within a group is critical: When an overwhelming majority seem ready to adopt a new idea, this convinces even the skeptics to go along. A surprising finding is that when people are working together doing the same thing in synchrony with others—e.g., rowing together, dancing together—our bodies release endorphins, natural opiates that give a pleasant high as a reward for working together.[5]

Similarly, business research has shown that this sort of engagement—repeated cooperative interactions among all members of the team—can improve the social welfare of the group,[6] and also promotes the trustworthy cooperative behavior conducive for successful business partnerships.[7] In microfinance banks such as Grameen Bank, which are now common all over the developing world, strong social engagement is key to their success because strong engagement increases the likelihood that loans will be repaid.[8]

Another interesting aspect of engagement in the online digital world is revealed in the results from a recent experiment using Facebook. The findings are rather straightforward—something our grandmothers could have predicted—and yet also quite telling about the power of engagement. During the 2010 U.S. congressional elections, a group of scientists from Facebook and the University of California–San Diego, led by James Fowler, conducted a large-scale experiment in which they sent "get out and vote" messages to 61 million Facebook users and analyzed the effects of different types of messages.

Some Facebook users received only a "get out and vote" message and the researchers found that these messages directly influenced political self-expression, information seeking, and the real-world voting behavior of millions of people. But the extent to which the message affected voting behavior was disappointingly small.

Other Facebook users received the "vote" messages and in addition saw the faces of friends who had already voted. Showing these familiar faces to users dramatically improved the effectiveness of the mobilization message. What our grandmothers would have known, though, was that nearly all the social influence occurred between close friends who had a face-to-face relationship.

Real-world friends are simply different from Facebook-only friends. The researchers found that close friends exerted about four times more influence on the total number of actual voters than the message itself. In fact, their results suggest that each act of voting on average generated an additional three votes as this behavior spread through the real-world, face-to-face social network.

Social Pressure

What is going on here? How is it that the face-to-face network was so much more effective at getting people to act compared to the Facebook message that started the spreading effect? And how could we make use of this effect to get people on the same page in other situations? The Facebook voting example suggests that information by itself is a rather weak motivator. On the other hand, both the ape troop and Bell Stars examples suggest that seeing members of our peer groups adopting a new idea provides a very strong motivation to join in and cooperate with others.

There is growing evidence that the power of engagement—direct, strong, positive interactions between people—is vital to promoting trustworthy, cooperative behavior. For example, in evolutionary biology, mechanisms such as direct and network reciprocity and group selection all work by exploiting the locality of interaction.[9] When people interact in small groups, the ability to punish or reward peers is very effective at promoting trusted cooperative behavior.[10]

Knowing that strong social ties mobilize people to act, how can we best leverage this tendency? Standard economic incentives miss the mark because they frame people as individual, rational actors rather than as social creatures influenced by social ties. Further, there is strong evidence that economic incentives don't work very well anyway.[11] But social physics tells us that there is another way: by providing incentives aimed at people's social networks rather than economic incentives or information packets that are aimed at changing the behavior of individual people.

As we reported in the *Nature* journal *Scientific Reports*, PhD student Ankur Mani, visiting Masdar faculty member Iyad Rahwan, and I have been able work out mathematically how best to motivate people by using social network incentives to increase their cooperation.[12] These incentives alter idea flow by creating social pressure, increasing the amount of interaction around specific, targeted ideas, and thus increasing the likelihood that people will incorporate those ideas into their behavior.

To test this theory in the real world, I decided to target the problem of increasing physical activity levels within the Friends and Family community of young families that I described in the last chapter. Increasing activity levels is a challenge in Boston's cold winter months, when people tend to stay inside more and generally become less active. This is, of course, bad for their overall health, and even worse, the habit of inactivity is sticky: People have a hard time resuming their previous levels of activity even when the weather improves. This is also a sort of "tragedy of the commons" problem, in which the unhealthy actions of the few can raise the health care costs of the entire community.

And so Nadav Ahrony and I deployed FunFit, a system of social network incentives within the ongoing Friends and Family study that encouraged people to remain active. Everyone in the Friends and Family study was assigned two buddies. Some individuals had buddies they interacted with a great deal, while some only had acquaintances. Since nearly everyone in the community was involved, each participant was also a buddy for someone else, and thus everyone had opportunities to be both a behavior-change target and a buddy.

As shown in Figure 5, the first step to deploy FunFit was to create clusters within the existing social network that were centered around each target person. The members of the cluster are called buddies (the light gray people in Figure 5), and they are given a small cash reward based on the behavior of the central target person (the dark gray people in Figure 5, labeled A and B) during the previous three days. This arrangement creates social pressure to be more active by providing incentives to the people who have the most interaction with the targets and rewarding them rather than the targets for improved behavior. In other words, our social network incentives promote engagement—repeated cooperative interaction among members of the team—around ideas about how to be more active.

For the entire community, we measured activity levels by using the accelerometer sensors embedded in their mobile phones. Unlike typical social science experiments, FunFit was conducted out in the real world, with all the complications of daily life. In addition, we collected hundreds of thousands of hours and hundreds of

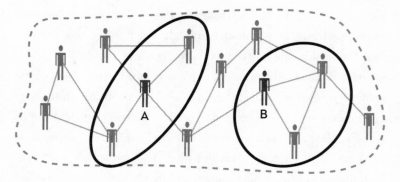

Figure 5. By confining incentives for good behavior to the local social network (the social ties in the dark ovals), social pressure is brought to bear on the target individual. This method works.

gigabytes of contextual data, so that we could later go back and see which factors had the greatest effect.

On average, it turned out that the social network incentive scheme worked almost four times more efficiently than a traditional individual-incentive market approach.[13] For the buddies who had the most interactions with their assigned target, the social network incentive worked almost eight times better than the standard market approach.[14]

And better yet, it stuck.[15] People who received social network incentives maintained their higher levels of activity even after the incentives disappeared. These small but focused social network incentives generated engagement around new, healthier habits of behavior by creating social pressure for behavior change in the community.

Given what we now know about the power of social ties, it is perhaps not too surprising about what postdoctoral students Erez Shmueli and Vivek Singh and I discovered: The number of direct interactions that people had with their buddies was an excellent predictor of how much their behavior would change.[16] In other words, the number of direct interactions is a very precise measure of the strength of the social pressure exerted between people.[17] Moreover, the number of interactions also predicted how well people maintained their new, healthier behavior after the experiment ended.

Similarly, the number of times people had direct interactions with each other gave a surprisingly accurate prediction of the trust they expressed in each other.[18] That is, the amount of direct interaction between two people predicts both the shared level of trust and the effectiveness of peer pressure.

The social physics approach to getting everyone to cooperate is to use social network incentives rather than to use individual market incentives or to provide additional information. That is, we focus on changing the connections between people rather than focusing on getting people individually to change their behavior. The logic here is clear: Since exchanges between people are of enormous value to the participants, we can leverage those exchanges to generate social pressure for change. Engagement—repeated cooperative interactions among members of the community—brings movement toward cooperative behavior.

Social network incentives act by generating social pressure around the problem of finding cooperative behaviors, and so people experiment with new behaviors to find ones that are better. The social pressure that is generated is a function of the cost of any mismatch between the behavior of the individuals, the value of the relationship, and the amount of interaction. This means that the most effective network incentives should be focused on the people who have the strongest social ties and the most interaction with others.

Digital Engagement

In our new, digitally connected, world, we now need to get people to work together using digital social media. Sometimes things go well and other times they don't seem to work out. How can we improve engagement in the digital world?

To understand how this might work, let's look at a variation on FunFit that we called Peer See. Here the idea was to duplicate the sort of conditions we saw in the Facebook voting experiment: that is, to use comparisons with peers to promote cooperation. In this Peer See experiment, we not only rewarded individuals for staying active (the standard economic incentive), but in addition we let them see online how their buddies were doing—a sort of social network incentive that works when people are in competition with each other.

We found that this Peer See approach, a sort of combined economic-social incentive scheme, was twice as effective as just rewarding people individually, without any social component (e.g., the standard economic incentive approach). The social pressure generated just by seeing what their buddies were doing doubled the effectiveness of the financial incentive.

With this example, we can now we can begin to understand why face-to-face relationships were so important in the Facebook voting example. Similar to the social incentives in the Peer See experiment, the knowledge that our face-to-face friends had already voted generated enough social pressure that it convinced people to vote. The Facebook message itself was relatively ineffective, but the few people it motivated to vote kicked off a cascade of voting among their face-to-face friends. Why mostly just face-to-face friends? Again, because social pressure depends on the strength of the social tie and the amount of interaction. Facebook friends don't really count in the same way (but our grandmothers already knew this).

For another example of combining social pressure and digital networks, consider the following energy conservation experiments, in which Ankur Mani, Iyad Rahwan, and I, and our colleagues Claire-Marie Loock, Thorston Staake, and Elgar Fleisch at the university ETH Zurich in Switzerland worked with the electric utility company to encourage electricity conservation among home owners throughout an entire region of their country.[19]

In the first experiment, home owners received social feedback on how much electricity they used relative to the average person. When the comparison was between the home owner and all other people in their country, virtually no savings resulted; people behaved the same. When the comparison was between them and people in their neighborhood, however, things worked better, showing that how closely they identified with the people in the comparison group mattered. This is a social network effect: Identification with a group of people increases both trust of group members and the social pressure that the group can exert.

These results suggested trying an approach based on social physics. So along with our colleagues at ETH, we next deployed a digital social network as part of the electric utilities' Web pages and gave people small rewards to encourage them to form local buddy groups. As in the FunFit experiment, this buddy network used social network incentives rather than standard economic incentives: When a person saved energy, then gift points were given to their buddies.

This social network incentive caused electricity consumption to drop by 17 percent, twice the best result seen in earlier energy conservation campaigns and more than four times more effective than the typical energy reduction campaign.[20] Just as in the FunFit experiment, behavior change was most effective when it leveraged the strength of the surrounding social ties.

The same combined digital and face-to-face dynamic happens with the use of social media inside companies. This pattern of engagement is of particular interest to organizations that are spread across continents and time zones, because in many situations the major mode of interaction between employees is digital social networks, e-mail, and short messaging. Because these media don't have all the social signals associated with face-to-face interactions, or even with audio interactions, companies have been disappointed to find that they are often much less effective at supporting the engagement required to develop an efficient work group (see the special topics boxes Digital Networks versus Face-to-Face [page 172] and Social Signals [page 131]).

Clearly there is a need to make digital networks more effective in the business world. To understand the situation better, PhD students Yves-Alexandre de Montjoye and Camelia Simoiu and I examined the growth and performance of more than one thousand companies' internal digital social networks.[21] We analyzed millions and millions of invitations, likes, and postings for each of these companies for an average of one year, looking for telltale patterns.

What we found was surprising: When the digital social network grows in bursts of engagement, the network ends up being far more effective than if it grows gradually. In companies in which people received a flurry of invitations to join the company's digital social network, they were much more likely join and use the network than they were in response to the same number of invitations spread out over time. But in companies where there were no bursts of engagement, few people joined, and things went from bad to worse. Just as the Bell Stars understood, until people see that there is a rush to adopt a new behavior, most group members will be reluctant to go along.

It didn't seem to matter if the bursts of activity came from the boss urging people to use it or from day-to-day deadlines. Because just as in the Facebook voting example, it was social pressure that really got people to work together; who invited whom to join and use the digital social network mattered the most. If the invitations were between people who already had a history of regular exchanges, and especially if those people were engaged with other people in the same work group, then their invitations were far more effective compared to invitations from other people with weaker social ties.

In fact, if anyone got three or more invitations to join the network within a half hour, and if those invitations were from people who were already engaged with them and their work group, then they were almost certain to join and give the digital social network a try. In contrast, even as many as twelve invitations within a half hour had relatively little effect if the invitations were from people who were not engaged with them or their work group.

If the use of a new digital tool is thought of as a change of habit, this pattern is exactly what would be expected. Recall my discussion of fast and slow thinking in Chapter 3: to effectively change a habit requires several examples of trusted peers successfully using or recommending a new idea within a short period of time. A rich social learning environment, made up of many examples from trusted peers, is required for people to adopt the habit of using a new social network within a company. But because most digital social media are asynchronous, it is often difficult to get this sort of repeated, frequent exposure. As in the Facebook voting example, it is more typical that the use of digital social media spreads through face-to-face networks rather than through digital networks only.

The data from these one thousand companies suggest that a good idea for driving the adoption of a new digital tool is by using social network incentives. For instance, one could reward people for how much their coworkers use the network to transact business with them. Such an incentive generates social pressure to use the network and might kick-start the process of creating new habits around use of the network.

The conclusion from this set of experiments is that engagement—repeated cooperative interactions—builds trust and increases the value of a relationship, which lays the groundwork for the social pressure needed to establish cooperative behaviors. In other words, engagement builds culture. Further, we have demonstrated that social network incentives accelerate this process and are often far more effective than using individual incentives.

Why don't companies rely more on social network incentives? Perhaps part of the reason is because social incentives have seemed to be fuzzy, vague, and simply "feel good" strategies rather than reliable tools of management. As a result, the social incentives managers typically use—e.g., employee of the month awards—are usually unconnected to real social relationships and so feel awkward and fake.

This can all change with social physics, however, because social physics provides a new, practical method that specifies how to create social incentives that establish more cooperative behaviors, and so improve everyone's situation. Social physics gives us new cost-benefit equations that work better than economic incentives and opens up new practical opportunities to promote cooperation.

Subjugation and Conflict

Adam Smith argued that the social fabric created by the exchange of goods, ideas, gifts, and favors guided capitalism to create solutions for the good of the community.[22] I agree. Communities are made of social ties, and without the constraint of social pressure provided by social ties, capitalism can become predatory. Social physics tells us that we must include not only economic exchanges, but also exchanges of information, ideas, and the creation of social norms in order to more fully explain human behavior.

Adam Smith's description of "good" capitalism outlined an ideal situation. He could imagine that social engagement almost always balanced economic forces, because he lived in a smaller world: Bourgeois residents in a city were more likely to know each other and thus be constrained by similar social norms and pressure to be good citizens. But his was also an era in which the poor were invisible and the lack of engagement between the rich and poor removed the social constraints on exchanges between these groups. Famously, this gave rise to the horrors and abuses of the first industrial age.

The same sort of disconnects can happen wherever there are

different ethnic, religious, or economic groups. In a recent paper in *Science*, May Lim, Richard Metzler, and Yaneer Bar-Yam showed that between-group violence is likely to happen when the communities are poorly integrated, when one group can dominate the other, and, in addition, when the political or geographic boundaries fail to match demographic borders.[23] Examples include the forced relocation of Native American tribes in the United States during the 1800s, the Catholic-Protestant clashes in Ireland, and repeated pogroms against Jews throughout Eurasia.

When there is a mismatch of this type, subjugation and persecution often follow. Majority groups have the power to define local rules, but if minority groups are large enough, then conflicts are likely to arise. With the right borders or with good integration, violence becomes unlikely.

How does this type of violence square with our results from the Friends and Family study in which we found that engagement builds trust? The answer is that the experiments I've described are in communities and social groups in which the vast majority of interactions are cooperative. If the majority of interactions were instead exploitative, then each interaction would serve to destroy trust. If someone gets ripped off every time they interact with a person from a different community, they will quickly come to distrust everyone in that community.

Because trust is the expectation of future cooperative behavior and is based on previous interactions, people seem to operate by what I call the reverse golden rule: Do unto others as they have done to you.[24] This is similar to the tit-for-tat strategy that is often seen in trust games, such as the classic prisoner's dilemma problem, but now applied as a general, default strategy.

Unfortunately, people quickly learn to apply this rule differently with peer groups versus with others; that is, trust peers and don't trust the others. This is why many people distrust politicians and lawyers as a group but not the specific politicians or lawyers that they know socially. This is what enables discrimination between groups, and even clan warfare. The dangers of systematic exploitation of one group by another highlights the importance of promoting cooperative interactions across different peer groups.

Rules of Engagement

From here on I will be using the term "engagement" to refer to the process in which the ongoing network of exchanges between people changes their behavior. As with the concept of exploration, there are three key things to remember about engagement:

> **Engagement requires interaction:** If people are to work together efficiently, there needs to be what is called network constraint: repeated interactions between all of the members of the group—not just between a leader and the members, or between the members and the entire group (as at a group meeting). The extent to which good network constraint has been achieved can be tested by asking if the people you talk to also talk to each other. If not, get them talking: We found that the number of direct interactions was a very good measure of the social pressure to adopt cooperative behaviors. Moreover, the number of interactions also predicted how well people maintained their new, more cooperative behaviors.

Engagement requires cooperation: Remember the Bell Stars: They pushed everyone on the team toward joint ownership of the group, involving everyone in goal setting, work activities, and getting credit for group accomplishments. These star performers promoted engagement within the team by making everyone feel part of the team, and they tried to reach sufficient consensus so that everyone would willingly go along with new ideas.

Building trust: Trust, by which I mean the expectation of future fair, cooperative exchanges, is built from the history of exchanges between people. Consequently, social networks have both history and momentum. As with cooperation, the number of direct cooperative interactions also gives a surprisingly accurate prediction of their trust. Social network pioneer Barry Wellman's suggestion that the number of telephone calls between two people is a good measure of their investments in the relationship—an investment often called social capital—seems exactly right.

In short, success at being part of a team depends on having continual engagement with the team network. People behave like players in a sports team, balancing individual ambition against social pressure to jointly develop behavior norms and patterns of trust and cooperation. In the next chapters, we will see that as a consequence the level of engagement is a strong predictor of team productivity and resilience across a wide spectrum of human activities.

Next Steps

What I have argued using the examples in these last three chapters is that idea flow, i.e., the spreading of new behaviors through a social network, may be conceptualized as exploration to harvest new ideas followed by engagement with peers to sift through those ideas and convert the good ideas into habits. Idea flow functions through social learning and social pressure to establish compatible norms of behavior. And finally, social network incentives, which alter the dynamics of idea flow, can be used to efficiently shape the spread of new behaviors.

The power of the surrounding flow of ideas over behavior seems central to human nature. In human tribal groups, decisions that affect the tribe as a whole are made in a social context and are determined by rich social cues of approval or disapproval;[25] they allow the group to weigh the preferences of all participants before a consensus is reached and enforced. Even ape troops decide group movements by a social consensus that is determined by social signaling.[26] Examples of groups creating and enforcing behavior norms range from the desperation of teenagers to fit in to the casual violence of gang members and rogue soldiers. When peers all adopt a new behavior, it is hard not to follow along.[27]

Some social scientists may ask what the fuss is all about. Aren't the experiments in the last three chapters just highlighting effects we already know about, such as homophily (birds of a feather flock together) and social learning (when in Rome we do as the Romans do)? Yes, but no one has really followed through on the computational effects of these well-known human behavior patterns: how these patterns of communication affect individual decision making and the fitness of the community. I have shown that these social universals produce a major increase in the collective intelligence of the community and in the ability of the community to act in a co-ordinated manner. Further, as we will see in the following sections of the book, these computational effects are central to the functioning of companies, cities, and society overall.

The following special topic box on the Mathematics of Social Influence gives a flavor of how to convert these ideas into equations that describe how our social fabric responds to new ideas and new incentives. By using these equations, we can reliably predict how an individual's behavior will change, and even what the work group performance or community outcome will be. More detail about the equations of social physics can be found in the Math appendix.

In the next sections of this book I will describe how these ideas and equations can be used to measure and manage corporations, cities, and even society in general. My hope is that these examples will give a real sense of both the potentials and dangers of living in our emerging hyperconnected society, and suggestions about what changes we must make in order to protect ourselves and to prosper.

THE MATHEMATICS
OF SOCIAL INFLUENCE

●

Most people don't speak math, so the main body of this book doesn't have any math. Unfortunately, this leads people to forget that social physics enables the building of predictive, mathematical models of human behavior that are being used to build better human organizations. So here is an English translation of part of the math, just so you can see what it is like.

For more than fifty years, social scientists have explored the question of who influences whom in social systems, but much of this research has been only qualitative or correlative. The challenge has been how to model social influence in a formal, mathematical way. An added complication has been that influence is often not directly observable, and thus must be inferred from individual-level behavioral signals.[28]

Our model of influence begins with a system of people C: Call it a company. Each person c from the first to the last (written as c = (1, . . . , C)) is at the beginning an independent actor, and what they are doing is normally hidden from casual observation—the ideas that drive their behavior are hidden in their heads. Let's write h to be the *hidden behavior ideas* of person c at time t as $h_t^{(c)}$. While we can't know directly what each person is thinking, their behavior gives us *observable signals* $O_t^{(c)}$, and the probability—$\text{Prob}(O_t^{(c)}|h_t^{(c)})$—of these signals depends on their hidden state, that is, on what is going on in their heads.[29]

Defining social influence in terms of state dependence—how one person's state impacts other people's states and vice versa—is an idea with deep roots,[30] and it lets us express social influence as the conditional probability between each person's hidden state $h_t^{(c)}$ at time t and the previous states of all people $h_{t-1}^{(1)}, \ldots, h_{t-1}^{(C)}$ at the previous time t-1. Consequently, the state $h_t^{(c)}$ of person c at time t is influenced by the state of all the other people at time t-1, and the conditional probability that person c is in state $h_t^{(c)}$ depending on their previous state at $t-1$ is

$$\text{Prob}(h_t^{(c)}|h_{t-1}^{(1)}, \ldots, h_{t-1}^{(C)}) \qquad (1)$$

The influence model breaks this overall "company state" into the influence each person c has on a particular other person c':

$$\text{Prob}(h_t^{(c')}|h_{t-1}^{(1)}, \ldots, h_{t-1}^{(C)})$$
$$= \sum_{c=(1, \ldots, C)} R^{c',c} \times \text{Prob}(h_t^{(c')}|h_{t-1}^{(c)}) \qquad (2)$$

Where the *influence matrix*, $R^{c',c}$, captures the influence strength of person c over c' and describes how influence spreads through the company's social network. The number of parameters in this model grows relatively slowly with increasing numbers of people and their internal states, making it easy to mathematically model "live" data and use it in real-time applications. Practically, this means we can determine the influence model parameters— influence, states, etc.—without knowing the social ties or learned behaviors beforehand by using an expectation maximization algorithm. Matlab code for estimating parameters and example problems are available at http://vismod.media.mit.edu/vismod/demos/influence-model.

This model accurately describes the behavior of investors in the eToro example. For the FunFit example, we add incentives that bias each person c, so that they are more likely to be in states that, via the influence matrix, influence their target c' to be in the desired behavior state. For instance, incentives might make c more likely to talk to c' about being more active, and as shown in the

FunFit experiment, the effectiveness of that action depends on the amount of interaction c has with c'.

As a consequence we can get a good estimate of social influence ($R^{c,c}$) by measuring the amount of interaction between c and c'. For almost all of the examples in this book, including the role of social influence on political views, purchasing behavior, and health choices, as well as productivity in small groups, departments within companies, and entire cities, we find that using measures of the amount of social interaction—both direct and indirect—in order to estimate social influence produces accurate estimates of future behavior.

A key question is how generally the estimated model parameters accurately represent real influence in human interactions. We have found that the model can accurately identify people's social roles—protagonist, attacker, supporter, neutral, and so on, in small groups, and in organizations, the model has allowed us to accurately map organizational relationships, clustering people into work groups and identifying group leaders.[31] And of course variations on the same basic model account for almost all of the examples in this book. Finally, a derivative of the model is currently used commercially to map the purchasing patterns of 100 million smartphone users (see http://www.sensenetworks.com, the Web site of a company I cofounded).

One of the most important consequences of this model is that it lets us take raw observations of behavior and gives us the social network parameters we need to get a numerical estimate of idea flow, which is the proportion of users who are likely to adopt a new idea introduced into the social network. Idea flow takes into account all the elements of the influence model: network structure, social influence strength, and individual susceptibility to new ideas.

In the eToro example, we found that the profitability of traders depended strongly on the rate of idea flow, providing us with the means to measure the quality of decision making within an organization or social network. In later chapters we will see that it also predicts productivity and creative output.

Finally, quantitative estimates of idea flow also lets us tune networks to perform better, because it gives us a way to predict the results of changing a network structure, influence strength, or individual characteristics.

PART TWO

Idea Machines

•

Collective Intelligence

HOW PATTERNS OF INTERACTION TRANSLATE INTO COLLECTIVE INTELLIGENCE

To continue building an understanding of how the physics of social interaction works, let's move on to examining interactions within smaller groups of people. Groups of people, as well as communities, also have a collective intelligence that is different from the individual intelligence of each group member. Moreover, this group intelligence is about as important a factor in predicting group performance as IQ is in predicting individual performance. This surprising finding, published in the journal *Science* by my colleagues Anita Woolley, Christopher Chabris, Nada Hashmi, and Tom Malone, and me, is based on research that examined the collective intelligence of groups by studying hundreds of small groups who were asked to perform a wide range of brainstorming, judgment and planning tasks, and to take IQ tests as a group.[1]

What is the basis of the collective intelligence we uncovered? Unexpectedly, we found that the factors most people usually think of as driving group performance—i.e., cohesion, motivation, and satisfaction—were not statistically significant. The largest factor in predicting group intelligence was the equality of conversational turn taking; groups where a few people dominated the conversation were less collectively intelligent than those with a more equal distribution of conversational turn taking. The second most important factor was the social intelligence of a group's members, as measured by their ability to read each other's social signals. Women tend to do better at reading social signals, so groups with more women tended to do better (see the Social Signals Special Topic Box [page 131]).

What was it that these women were doing to improve the performance of the group? The social physics viewpoint suggests that it should have something to do with idea flow within the group. Fortunately, during the various group tasks, people in many of the groups were being monitored by my research group's sociometric badges. Postdoctoral student Wen Dong and I later analyzed the sociometer data from these experiments in order to measure the patterns of the idea flows.[2]

The badges used in this experiment, and in other research studies from my lab, produce detailed, quantitative measures of how people interact. Typical variables measured include: the tone of voice used; whether people face one another while talking; how much they gesture; and how much they talk, listen, and interrupt each other. By combining data from individuals within a team and comparing it with performance data, we can identify the interaction patterns that make for successful teamwork (see the Reality Mining appendix).

What these sociometric data showed was that the pattern of idea flow by itself was more important to group performance than all other factors and, in fact, was as important as all other factors taken together. Think about it: Individual intelligence, personality, skill, and everything else together mattered less than the pattern of idea flow.

Wen and I found that three simple patterns accounted for approximately 50 percent of the variation in performance across groups and tasks. The characteristics typical of the highest-performing groups included: 1) a large number of ideas: many very short contributions rather than a few long ones; 2) dense interactions: a continuous, overlapping cycling between making contributions and very short (less than one second) responsive comments (such as "good," "that's right," "what?" etc.) that serve to validate or invalidate the ideas and build consensus; and 3) diversity of ideas: everyone within a group contributing ideas and reactions, with similar levels of turn taking among the participants.

These patterns, which are illustrated in Figure 6, are pretty

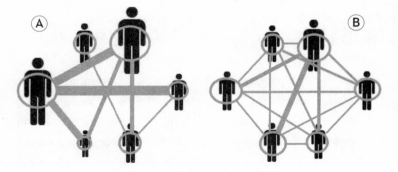

Figure 6: (a) an unproductive pattern of interaction, (b) a good pattern of interaction.

much the same as we've seen in Chapters 2, 3, and 4, and what I see in the most creative people I know: exploration for idea discovery and engagement to select the best ideas and make sure everyone is on the same page. And, just as before, diversity of ideas is a critical variable.

One exception to using these patterns of interaction as a guide is performance in times of stress. When the decision needs to be made *now*, there may be no time to get all the ideas out and discuss them. A second exception is when the group has a hard time working together and emotions are high; then a leader may have to play the role of facilitator and frequently intervene between contributions by others. These interventions should be as short as possible, though, to leave air time for new ideas.

The sociometric data from these small working groups highlight that teams are operating as idea-processing machines in which the pattern of idea flow is the driving factor in performance. Group performance in the *Science* study depended upon how good the group members were at harvesting ideas from all of the participants and eliciting reactions to each new one. It seems that what the women and the other socially intelligent participants in our collective intelligence experiment may have been doing was enabling better idea flow by guiding the group toward briefer presentations of more ideas, encouraging responses, and ensuring that everyone contributed equally.

How can the pattern of idea flow be as important as all other factors taken together? For the answer to this question, we look to our ancient ancestors. Language is a relative newcomer, evolutionarily speaking, and was likely layered upon older signaling mechanisms for dominance, interest, agreement, etc., in order to find resources, make decisions, and coordinate action. Today these ancient patterns of interaction still shape how we make decisions and coordinate among ourselves.

Consider how our ancient ancestors may have approached problem solving. One can imagine ancient humans sitting around a campfire making suggestions or relating observations, with other participants responding by signaling their level of interest or approval using head nods, gestures, or vocal signals. To determine if a particular idea is approved by the group, a member needs only "add up" the responsive signaling to determine if most group members are in agreement.

Early human groups needed to pool ideas in order to solve shared problems, much as ape groups are observed pooling ideas today. Animal behavior research supports the idea that this is what ape troops, and even bee colonies, do when deciding about group actions. Our sociometric badge data demonstrate that this is also what happens in modern group problem-solving sessions. The back-channel "ums" and "OKs" that greet new ideas in today's conference rooms preserve and leverage these ancient mechanisms for sorting through alternative ideas.[3]

The important conclusion to take away from this *Science* paper is that groups have a collective intelligence that is mostly independent of the intelligence of the individual participants. This group problem-solving ability, which is greater than our individual abilities, emerges from the connections between the individuals. In particular, a pattern of interactions that supports the pooling of a diverse set of ideas from everyone, combined with an efficient winnowing process to establish a consensus, seems to form its core. Have we evolved to function better as group minds rather than as individuals?[4]

Measuring What You Manage

Our study on the collective intelligence of groups revealed teams to be functioning as idea-processing machines in which the pattern of interactions facilitates a type of data mining of ideas. Simply measuring a group's interactions pattern allows us to accurately predict the eventual productivity of the group.

I see this in companies too. Some companies give the impression of a well-oiled machine, or of a complex puzzle in which all of the parts fit together perfectly. The natural question is: Can a company's performance be measured by looking at just the patterns of interactions? Are organizations—companies or governments—also operating as idea machines, harvesting and spreading ideas primarily through individual interactions?

In the workplace we don't just sit around a table, as in a laboratory experiment. Instead we move around during the workday and talk to people at their desks, in the hall, at lunch, and in small informal gatherings around the coffee station or printer. So I began to instrument actual work groups using sociometric badges in order to measure face-to-face interaction patterns in a variety of real-world settings.

My former PhD students Taemie Kim, Daniel Olguín Olguín, and Ben Waber, who are now colleagues at our spin-off company, Sociometric Solutions, and I used these sociometric badges to study a variety of workplaces, including creative and research teams in companies, post-op wards in hospitals, plain old back-room operations, and call centers.[5] In order to understand the total pattern of interactions within an organization it is vital to capture all data from the media in use, including e-mail, IM, and others. After instrumenting all these channels of communication, we then examined the interaction patterns within both high and low productivity groups.

As I reported in my *Harvard Business Review* article "The New Science of Building Great Teams," my research group and I have collected hundreds of gigabytes of data from dozens of workplaces.[6] What we found was that the patterns of face-to-face engagement and exploration within corporations were often the largest factors in both productivity and creative output. In the next sections of this chapter I will explain how these patterns influence work output and how companies can make use of this insight.

Productivity

As a first example, let's examine data from a call center. Call centers are unusual in that they are very highly instrumented and keep track of pretty much everything. Call center operators often try to minimize the amount of talking among employees because operations are so routine and standardized; they feel that employees have little to learn from each other. This can take various forms, but a common one is to give employees staggered break times.

In 2008, we had just begun a relationship with Bank of America, and I thought that the tightly managed environment of a bank's call center could be an acid test of the hypothesis that idea flow among employees was a prime factor in productivity. I proposed to Bank of America that my research group would measure the employee interaction patterns and then execute a simple intervention to see if we could improve idea flow.

And so we began a two-part study at a call center with over three thousand employees. During the first phase, my group targeted four teams, each consisting of around twenty employees.

They were instructed to wear the sociometric badges all day while they were at the call center for a period of six weeks. In total, some tens of gigabytes of behavior data were collected.

In this call center, the most important measure of productivity is something known as average call handle time (AHT), because this factor dominates the dollar cost of running a call center. For example, an intervention that reduced AHT by only 5 percent at this single call center would save the company roughly $1 million per year.

When we analyzed the large data set we collected, we found that the most important factors for predicting productivity were the overall amount of interaction and the level of engagement (the extent to which everyone is in the loop). Together these two factors predicted almost one third of the variations in dollar productivity between groups.

This example illustrates how performance varies with idea flow within this call center work group, and it is similar to how we analyzed the eToro trading network (see Figure 3 in Chapter 2). Again, once we have this diagram of idea flow versus performance, we can then tune the network to improve performance.

In this case, I proposed to management that we could change the coffee break structure of the call center. The standard pattern of coffee breaks for these employees, as in many call centers, was to give a break to one person at time. Because this organization has so many employees, however, it was just as easy to shift call loads between teams as within teams. That meant that it was just as easy to give breaks to entire teams at the same time as it was to give breaks to individuals one at a time. In order to increase the amount of informal interaction and proportion of engagement among employees, I suggested giving all the employees on a team a break at the same time.

Allowing employees to mix more during breaks raised the amount of interaction within each work group in the center as well as the level of engagement between employees. The AHT decreased sharply, which means that the employees were much more productive, demonstrating the strong link between interaction patterns and productivity. As a result of this simple change, the call center management converted the break structures of *all* their call centers to this new system and forecast a $15 million per year productivity increase.

This case study clearly demonstrated that the level of face-to-face engagement has a major effect on productivity. Does this same effect apply to other work situations as well? To examine this, we used our sociometric badges to instrument a typical white-collar, back-room operation that configured IT solutions to support the sales staff. In this study, our measurements focused on a sales support team that consisted of twenty-eight employees, with twenty-three participating in the study. Our sociometric badges were deployed in this Chicago-area data-server sales firm for a period of one month (twenty working days), collecting roughly a billion measurements about who talked to whom, their body language, and even their tone of voice. In total, nineteen hundred hours of data were collected, with a median of eighty hours per employee.[7] (For data, papers, and additional detail, see http//realitycommons .media.mit.edu).

Our analysis examined employee behavior during each sales-support task. Employees in the department were assigned a computer system configuration task in a first-come, first-served fashion.

Each employee later submitted the completed configuration as well as the price back to the salesman, and then that employee was placed at the back of the queue for a new task assignment. The exact start and end time of the task was logged, allowing us to calculate the exact dollar productivity of each employee for each task.

What we found was that engagement was the central predictor of productivity. Remember that engagement is defined as idea flow within a work group; in this case, it was measured by computing the degree to which the work group members that each employee talks to also talk to each other. Controlling for all other factors, including length of employment and gender, workers whose measured engagement was in the top third had productivity that measured more than 10 percent higher when compared to the typical employee.

Thus, in this white-collar operation, we again see that the concept of idea flow is key to understanding the relationship between productivity and interaction patterns. It appears that being in the loop allows employees to learn tricks of the trade—the kind of tacit, detailed experience that separates novices from experts—and is what keeps the idea machine efficiently ticking along.

Creativity

Not only do patterns of interactions have a major effect on productivity, but they also influence even our most sophisticated creative abilities. The sociometric data that my research group and I have gathered from many different organizations show that creative output depends strongly on two processes: idea discovery (exploration) and the integration of those ideas into new behaviors (engagement). In research labs and in design shops, the difference between low-creative groups and high-creative groups is their pattern of face-to-face exploration outside of the group, together with their engagement within the group.

Although both exploration and engagement are crucial to creative output, each places different and conflicting demands on the pattern of interactions. The solution suggested by other social species, such as ape troops and bee colonies, is to alternate between exploration for idea discovery and engagement with others for behavior change.[8]

This ancient mechanism for combining resource discovery with group decision making is one that drives many organizations, both human and nonhuman. The humble honeybee, for instance, has much to tell us about good patterns of social interaction. It is common knowledge that worker bees explore for good food sources and then return to the hive and do a waggle dance to signal the distance and direction of the food. This special dance serves to encourage other workers to change their behavior and visit the new food source.

Less well-known, though, is that bees use this same mechanism as the basis for group decision making. One of the most important choices made by a bee colony is where to locate a nest, and bees use a kind of idea machine to make this decision. The colony sends out a small number of scouts to explore the local environment. When they return to the hive, scouts that have found promising sites communicate their discovery to the other bees by doing an intense, active dance. This causes other bees to change their behavior and accompany them back to check out the same site. When those scouts return, they recruit more bees to the site with their dancing, and the cycle continues until, eventually, so many scouts are signaling in favor of the best site that a tipping point is reached, and the hive moves en masse.

The bees' decision-making process highlights the oscillation between exploration as a method of resource discovery and engagement as a method of spreading a new behavior throughout the peer community. As we will see, these two processes are also crucial to human organizations. Each process has different requirements. The solution suggested by the bees is to alternate between a star-shaped network that is best for exploration and a cohesive, richly connected network that is best for engagement, idea integration, and behavior change. Networks—whether apian or human—that vary their interaction structure as needed are able to shape idea flow to optimize both exploration and engagement.[9]

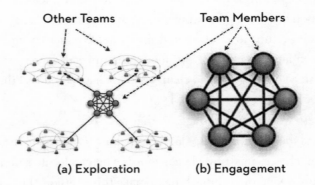

Other Teams   **Team Members**

(a) Exploration   (b) Engagement

Figure 7: Exploration and engagement networks.
(a) Exploration is when team members interact with other teams.
(b) Engagement is when they interact with each other.

In the typical corporate exploration pattern, workers try to reach out and interact with different teams, forming a star-shaped network; see Figure 7(a). In general, this facilitates the discovery of new, useful ideas by generating idea flow from outside the team. In the typical engagement pattern, workers adopt a densely interconnected pattern in which most interactions are with other team members. They have lunch and coffee with members of the team, they encourage friendships within the team, they make sure that shy members join in, and so forth: The idea is to get everyone talking to everyone else. This generates high idea flow within a team, thereby facilitating both the vetting of new ideas and their integration into the norms and habits of the team; see Figure 7(b).

Qualitatively, this is what the Bell Stars study of Chapters 2 and 3 found: Star performers became familiar with different perspectives on their work. Senior management, customers, sales, and manufacturing groups all have different views, and the combination of their ideas with those already in their work group were a major source of useful creative thinking. The difference today, of course, is that with sociometric badges we can now actually measure this exploration and ensure that it is both frequent enough and sufficiently diverse.

In order to verify that the pattern of alternating exploration and engagement correlates with creative output, my students and I, working with Peter Gloor and his collaborators, used our sociometric badges to measure the interaction patterns within the marketing division of a German bank. We instrumented a group of twenty-two employees (distributed into five teams) in the bank's marketing division for a period of one month (twenty working days). Each employee was instructed to wear a sociometric badge every day, and in total we collected twenty-two hundred hours of data (one hundred hours per employee). We also monitored e-mail traffic and logged 880 reciprocal e-mails.[10]

While analyzing these data, we found clear evidence of employee groups varying their interaction structure over time, shap-

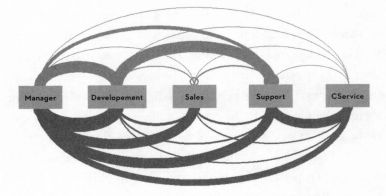

Figure 8. Interaction patterns during one day in the marketing division of a German bank. The thickness of the arc between two groups represents the amount of face-to-face interaction (light gray, on the top) or e-mail interaction (dark gray, on the bottom).

ing idea flow in a way that oscillated between exploration and engagement. As an example, Figure 8 illustrates one day of their interactions. The dark gray arcs on the bottom show the volume of e-mail between groups, while the light gray arcs on the top show the amount of face-to-face interaction.

Analysis of the data showed that teams charged with creating new marketing campaigns oscillated between patterns of exploration for discovery of new ideas and engagement for integration of these ideas into the team's behavior. This pattern is good at generating idea flow into the creative teams. In contrast, teams charged with production showed little oscillation, with members engaging mostly with other team members. Consequently, there was little flow of new ideas into those teams.

While studying this company, we also discovered a black hole for idea flow: people rarely talked face-to-face with the customer service group (labeled "CService" in Figure 8). By changing the seating plan the bank was able to make sure that everyone, even the previously orphaned customer service group, was in the loop. As a result of this simple change, some of the coordination problems that this department had previously experienced—for instance, launching ad campaigns that overwhelmed the customer service group—were now much improved.

But is this pattern of oscillation between exploration and engagement really the driver of creative output? To explore this question further, I studied the networking patterns within 330,000 hours of interaction data from ninety-four people at MIT, an experiment known as the Reality Mining Study conducted by graduate student Nathan Eagle and me.[11] The data were collected by smartphones that the participants carried, which allowed us to measure the face-to-face interaction patterns of various research groups within the Media Laboratory.

Graduate student Wen Dong and I discovered that these groups had also rated their creative output in a lab-wide human resources survey. When we compared the creativity ratings to the interaction patterns, what Wen and I found was that the teams that showed more variations in the shapes of their social networks also rated themselves as having more creative output.[12] In other words, greater oscillation between patterns of exploration and engagement within these social networks correlated with creative productivity, at least as judged by the people in the networks (see the Reality Mining appendix).

While these results tell us more about how the exploration/engagement cycle is correlated with subjective creative output, ideally we would have stronger evidence that this pattern is predictive of objective creative output. Unfortunately, objective measures of creative output are hard to find; who can say what is truly creative? But perhaps the best available measure is the KEYS creativity assessment tools developed by Professor Teresa Amabile at Harvard.[13] The KEYS scale is widely recognized as the gold standard for measuring team creativity and innovation within organizational work environments.

In her PhD thesis, Pia Tripathi, working with her thesis adviser, Win Burleson, and me, used the sociometric badges to study two R&D labs within the United States.[14] She gave them to two seven-member groups for eleven and fifteen days, respectively. The KEYS team survey was used to quantify both a self-rated and an expert-rated creativity assessment. The daily values of these two scores were then split into high and low groups and used to categorize creative versus noncreative days.

Analysis of the KEYS team-creativity data showed that people within these groups had both more engagement and more exploration during creative days than during noncreative days. In fact, a simple combination of the engagement and exploration measures was able to predict which days were the most creative with 87.5 percent accuracy.

And so once again we see that the pattern of alternating engagement and exploration promotes greater creative output. The exploration phase is ideal for bringing new ideas into the group, and the engagement phase is best for obtaining consensus about the idea. To use Herb Simon's phrasing, if there is consensus about an idea, it is then integrated into the team's store of "action habits" to use for their fast thinking. Put another way, the process of oscillation between exploration and engagement appears to increase creative output by building up a more diverse store of experiences that can be drawn on as examples.

The connection between creative output and diversity of experiences seems to be due to the power of unconscious cognition.

There is considerable evidence in the scientific literature showing that unconscious cognition can be more effective than conscious cognition for solving complex problems.[15] Our fast thinking seems to work best when our more logical, slow-thinking minds aren't interfering, such as during sleep or when we are turning an idea over in the back of our minds. Because fast thinking uses associations rather than logic, it can make intuitive leaps more easily by finding creative analogies. It can take the experience of a new situation, let it soak in for a while, and then by association produce an array of analogous actions. In contrast, our attentive, slow-thinking mode provides insight into our actions, helping us detect problems and work through alternate plans of action.

Improving Idea Flow

In this chapter, we have seen how patterns of idea flow affect the collective intelligence of work groups and organizations. In particular, I have focused on understanding how good idea flow can improve decision making, productivity, and even creative output. In studies of dozens of organizations, I have found that the number of opportunities for social learning, usually through informal face-to-face interactions among peer employees, is often the largest single factor in company productivity. In our research, we assess how much opportunity there is for social learning through the measurement of group engagement: Do the people one person talks to also talk to one another? How tightly woven and interconnected are peer networks?[16]

As a consequence of this relationship between opportunities for social learning and productivity, simple tricks to improve social learning often have enormous payoffs. As we have seen, in one case a simple change in the coffee break timing allowed employees to talk more easily with each other, with the result that productivity improved enough to save the company $15 million per year. In another company, the simplest way to increase workers' productivity was to make the company's lunch tables longer, thus forcing people who didn't know each other to eat together.[17] In the next chapter I will show how visualization of patterns of interaction can be used to improve idea flow by making everyone aware of the current patterns and then obtaining group agreement about how to improve them.

These examples underscore the extent to which our behavior comes from social learning and is a major reason that engagement with our local peers is vitally important. With increased engagement comes an increase in opportunities for social learning, for sharing vital resources such as tacit operational knowledge and successful work habits. In other words, many of the important ideas about how to be successful and productive at a job are likely to be found around the coffee pot or water cooler.

Shaping Organizations

SOCIAL INTELLIGENCE THROUGH VISUALIZATION OF INTERACTION PATTERNS

The social physics view of organizations focuses on patterns of interaction acting as a kind of "idea machine" to carry out the necessary tasks of idea discovery, integration, and decision making. Leaders can increase its performance by promoting healthy patterns of interaction within their organizations (including both direct interactions, such as conversations, and indirect interactions, such as overhearing or observing). This stands in contrast to a focus on the individuals in an organization or the specific content of information being spread. Instead, when we think of our organizations as idea-processing machines that harvest and spread ideas primarily through individual interactions, then it is obvious that we must establish healthy patterns of idea flow.

In studies of more than two dozen organizations I have found that interaction patterns within them typically account for almost *half* of all the performance variation between high- and low-performing groups.[1] This makes the pattern of idea flow the single biggest performance factor that can be shaped by leadership, and yet today there isn't a single organization in the world that keeps track of both face-to-face and electronic interaction patterns. And, as we all know, what isn't measured can't be managed.

In my *Harvard Business Review* article "The New Science of Building Great Teams" I argued that moving from a management that uses org charts to a management that monitors idea flow requires a shift away from the individual talent approach to managing organizations and a move toward shaping interaction patterns to achieve better collective intelligence.[2] By moving away from a static org chart to a focus on the real interaction network, we can bring everyone into the loop, to make it more likely that good ideas will turn into coordinated behaviors.[3]

Perhaps the first step to achieving good idea flow is to make people aware of their patterns of interaction. Unfortunately, it is usually difficult, if not impossible, for people to be aware of them. How can anyone know about face-to-face conversations that take place in the hall unless they're there? Or that one person learned how to operate the copier by watching another person use it?

The obvious idea is that by making the patterns of interaction visible to everyone, everyone can work together to create better patterns of idea flow. To this end, my research group has built interaction maps—tools for measuring and giving feedback about the patterns of interaction throughout a group or organization—so that members can begin to understand how ideas flow between individuals and within work groups. The goal is to increase the social intelligence of both work groups and the entire organization, and so increase their performance.

Once people can actually see the patterns of interaction, then they can begin to discuss how best to manage them. The discussion about which patterns need to be reinforced and which need to be reduced ideally results in a shared understanding of what needs to be changed. This shared understanding then produces social pressure to adopt the agreed-upon patterns.

When we instrument a typical organization in order to visualize interaction patterns, both managers and employees wear our specially designed sociometric badges (see the Reality Mining appendix for more detail). We then give everyone graphic feedback about their patterns in the form of a dashboard that can be pulled up on their computer screens or printed out to facilitate group discussion. The feedback can be in real time or (more commonly) delivered at the beginning of the next day.

The most useful visualizations convey the levels of engagement and exploration within the organization, since these are the two main patterns that are characteristic of healthy idea flow. In personal terms, the notion of engagement means that if the people you talk to also talk to each other, then you are in the loop and in good shape. We have found that engagement levels predict up to half of the variation in group productivity, independent of content, personality, or other factors. Exploration is how much the members of a group bring in new ideas from outside; that in turn predicts both innovation and creative output. Because innovation is the most important driver of long-term performance, it is important that managers encourage exploration for new ideas by helping employees establish diverse connections between people.

Engagement

Good idea flow is difficult in some kinds of groups, for example, in both widely dispersed and mixed-language groups. To address this common problem, my MIT research group has developed methods of providing real-time graphic feedback about engagement to promote higher performance within many types of work groups. The goal is to have people use these real-time displays to provide them with the sort of social intelligence needed to foster better interaction patterns, thus leading to higher productivity and creative output.

The tool shown in Figure 9, developed by Taemie Kim and me during her PhD work, is called the Meeting Mediator, and as shown in (a), it has two main components: a sociometric badge to capture turn-taking behavior and a mobile telephone to visualize the group's interactions.[4] In the Meeting Mediator system, the ball in the middle becomes bright green when there is high engagement within the group.[5] When everyone is contributing more or less equally, the pattern of interaction is healthy, and, as in (b), the ball moves to the center of the screen. When one person is dominating the conversation, as in (c), the ball becomes pale and moves toward the person who is speaking too much. This phone visualization provides real-time feedback to encourage balanced participation as well as high engagement within the group.

One advantage of this feedback system is that it is simple enough to be effective even when people don't pay conscious attention to the display. We have found that the Meeting Mediator is particularly effective for geographically distributed groups, often bringing their performance and even their level of trust up to the level of face-to-face groups.[6] As we saw in Chapter 4, high levels of trust are the bedrock on which flexible cooperation is built.

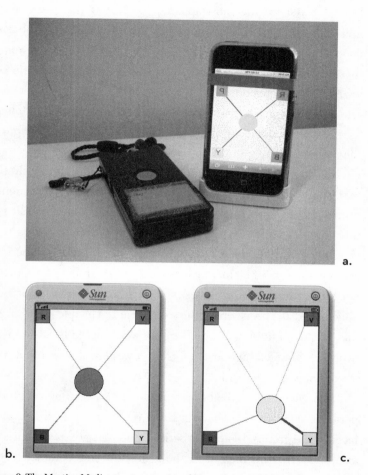

Figure 9: The Meeting Mediator system consists of (a) a sociometric badge (left) to record the interaction patterns of groups and a mobile phone (right) to display them as real-time feedback. When team engagement is high, as in (b), the ball in the center becomes dark, and when engagement is low, as in (c) the ball becomes pale and white. When the pattern of interaction is healthy and everyone is contributing equally, as in (b), the ball moves to the middle of the screen. When one person is dominating the conversation, as in (c), the ball moves over toward that person.

When geographically distributed groups use the Meeting Mediator system, the most apparent change is that there are a greater number of contributions per minute and the number of contributions by each member is more similar. In other words, there are lots of short contributions and everyone in the group engages in the discussion, with no one person dominating. When we remember the recipe for collective intelligence discussed in Chapter 5, we would expect that the Meeting Mediator would improve the productivity of groups.

Indeed, this change in group behavior produces exactly the performance increase we would expect. In laboratory experiments, distributed groups using the Meeting Mediator not only had significantly higher levels of cooperation, but they were indistinguishable from those found in face-to-face groups. Moreover, when we asked people about factors such as trust and their feelings of membership, we found the same result. Significantly, when researchers only looked at a transcript of the words that the people had spoken, they could not figure out which groups would actually act cooperatively or which felt high levels of trust. It was engagement that mattered, not what they said.

The Meeting Mediator system also helps distributed groups perform better at idea sharing, bringing them up to the level of face-to-face groups. In another series of laboratory experiments, we examined this by measuring how quickly and completely the groups managed to get out all the critical ideas within a test problem. These data showed that it was their similarity of engagement, especially a similar number of contributions per group member, that mattered most. Once again, this is consistent with the results of the collective intelligence experiments described in Chapter 5.

Surprisingly, this similarity effect was not limited to just the number of contributions in a discussion but also extended to their body language—even though the participants in these distributed groups couldn't see each other. In fact, the higher the performance of the group, the more people shared a common rhythm, including body movement, speech, and tone of voice. The best performing groups were in sync, literally moving in synchrony with each other.[7]

Similar visualization feedback also improves performance in mixed-language groups. For instance, in one demonstration project we deployed the sociometric badges at a leadership forum held in Tokyo, Japan. The forum brought together twenty students from the United States, mostly from universities in the greater Boston area, and twenty students from Japan, mostly from universities in the Tokyo area. Working in groups of six to eight people, the students agreed to take part in a training project that focused on creative engineering and required the cooperation of all group members. The student participants wore the badges during all working hours for a total of seven working days.

Worried that cultural and language barriers would damage the performance of these working groups, we wanted to encourage them to become more interactive and integrated into a single team. To accomplish this, we used the sociometric badges to measure their communication patterns each day, as they were working together. Then, at the end of each day, we provided the team members with a sheet of paper that included a visualization of the group communication patterns.

Figure 10 illustrates a typical team's face-to-face interaction patterns as captured by the sociometric badges at the beginning of the weeklong interaction.[8] The size of each circle represents the amount of time a person participated in the conversation and the thickness of the connecting line shows how much two people

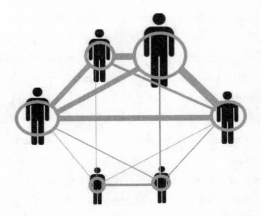

Figure 10: One team's face-to-face interaction patterns as captured by the sociometric badges, at the beginning of one week of interaction. The size of each figure represents the amount of time a person communicated, and the thickness of the link shows how much two people communicated.

talked. In this particular example, the two students at the bottom are Japanese, the rest are American.

At the beginning of the week, both the American and Japanese students interacted mostly among themselves. This poor level of group integration was in part because the group discussion was in English, so the Japanese participants faced language handicaps and cultural differences. By the end of the week, however, the team appeared fully integrated, and the overall pattern of interaction had improved dramatically. In a debriefing at the end of the week, the participants credited the sociometric feedback for helping them form more integrated, higher-performing teams.

Exploration

As we discussed in Chapter 5, creative output is critically dependent on exploration. Unfortunately, it is hard to be aware of a team's pattern of exploration, in part because it is usually done individually and not in a group. Because it is hard to see, it is also difficult to create organizational habits that support it. As a consequence, finding a way to visualize a group's pattern of exploration is extremely important to supporting good idea flow.

While the mathematical measure of idea flow between a work group and people outside it is probably the best way to measure exploration, we have found that it is usually adequate to simply count the number of outside interactions.[9] In other words, including complexities such as feedback loops or structural holes in the social network is only necessary in specific examples when we expect that these sort of network structures will cause problems.

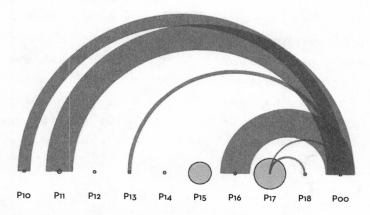

Figure 11: An exploration dashboard (courtesy of Sociometric Solutions).

An example of a display of group exploration behavior is shown in Figure 11.[10] In this display we see the pattern of interaction among groups (shown as gray arcs between the groups P00–P18) and the pattern of interaction within groups (shown as gray circles along the bottom). For intergroup interaction, the width of each arc shows the amount, while the size of the circle shows how much interaction there is within each group. As can be seen, the management group (P00) is well engaged with some of the groups but not at all engaged with others. Even worse, none of the groups have significant amounts of group-to-group interaction. Just as important, except for groups P15 and P17, all of them have virtually no within group interaction. This looks like a case of traditional top-down management at its worst.

As we discussed in the previous chapter, organizations with such poor patterns of exploration often find themselves stuck in old patterns of behavior. In addition, they are prone to major disconnects in what different work groups think they should be doing. For instance, recall the case of the German bank we studied in Chapter 5: The fact that no one talked to customer service meant that ad campaigns were often designed without considering whether or not customer service would be able to carry through the duties assigned to them. Managers need to visualize the patterns of communication using dashboards like those shown in Figure 10 (for engagement) and Figure 11 (for exploration) and take steps to make sure that ideas flow within and between all of their work groups.

Diversity

A key social intelligence problem is to know when there's enough diversity in the ideas that have been harvested. Whenever a social network has many loops, so that the same ideas circulate around and around, or the external channels of communication that help drive people's explorations are too similar, then there is insufficient diversity in the idea flow. There are three basic ways to detect and deal with this problem.

The first method might be called the bookie solution. Bernardo Huberman's research group at Hewlett-Packard (HP) has developed a scheme that first asks each person what they thought everyone else was going to say.[11] This "common knowledge" is then discounted, since it is obviously being counted more than once. This method has proven quite useful in idea markets, where people bet on the outcomes of election campaigns, movie box office returns, etc.

A second way to tackle this kind of echo chamber problem was invented by Drazen Prelec at MIT, who came up with what he calls Bayesian truth serum, which is a way of figuring out who has genuinely new bits of information that might make a difference.[12] One might also call this the wise guys solution to the problem of insufficient diversity in idea flow.

In the wise guys method, we look for individuals who can accurately predict how other people will act but whose own behavior is different. The logic is that if a person can predict other people's actions, then they already know the common knowledge. But if their behavior is also different from everyone else's, then they must know something the others don't. The behavior of such wise guys, then, can be counted as an independent bit of information.

A third solution for echo chamber problems is one that I came up with: To estimate the amount of social influence between people, keep track of the dependencies between people's ideas and their behaviors.[13] For instance, people who regularly have similar opinions probably have similar sources of information, so opinions by such birds of a feather can't count as independent. This typically happens within tight social groups, since the members of such groups often share information and there may be social pressure to hold the same opinions. By paying attention to idea flow within the network, we can discount these effects, allowing us to integrate opinions that are more likely to be truly independent.

In most practical applications, I have found that this third method, estimating the amount of social influence between people, is the easiest and works quite well. A practical method of accomplishing this for large groups is by the use of the influence model described in the Math appendix. For complex problems with many trade-offs, however, the wise guys solution may be best. Nothing beats having private information sources or successful new strategies.

Social Intelligence

In our *Science* paper characterizing the intelligence of groups, we found that group members with higher social intelligence enhanced the performance of the overall group across a wide range of tasks. What we have found in other studies is that the social intelligence within a group can be increased by providing visual feedback of their interaction patterns. This feedback improves the group's pattern of interactions, and this, in turn, improves the group's objective performance.[14]

In the preceding section we explored the advantages that can be derived from the visualization of and discussion about idea flow within groups. This feedback is a sort of computer-aided social intelligence that raises the performance of work groups by producing social pressure to improve the work group's patterns of interaction. What other ways can we use the ideas of social intelligence, social pressure, or social network incentives to improve idea flow within work groups?

One of the most common ways to grow a high-performance

culture is through the personal influence of leadership. Effective leaders usually have a sort of practical charisma: By being energetic and systematically engaging with others, they can help grow the interaction patterns of their organization in the right direction. Rather than dominating discussions within the group, they can encourage good patterns of idea flow.

This practical charisma is illustrated by a study that PhD student Daniel Olguín Olguín and I did with executives attending an intensive one-week executive education class at MIT in which the final project was pitching a business plan.[15] This time we used the sociometric badges to observe the executives during a mixer on the first evening of the course. Perhaps to the chagrin of the course organizers, we found that the executives' social styles at the precourse mixer were extremely predictive of how well their teams' business plans would be judged at the end of the course.

The most successful style is what I call the charismatic connector. These people circulated actively through the crowd and engaged people in short, high-energy conversations, acting rather like a bee harvesting pollen. We found that the more of these charismatic connectors a given team had among its members, the better the team performance was judged during the business plan contest at the end of the week. It seems that teams whose social style is dominated by these charismatic connectors may have had more evenhanded turn taking and high levels of engagement, which is the recipe for collective intelligence.

The charismatic connectors are not just extroverts or life of the party types. Rather, they are genuinely interested in everyone and everything. I think their real interest is in idea flow, although probably few would describe their interest that way. They tend to drive conversations, asking about what is happening in people's lives, how their projects are doing, how they are addressing problems, etc. The consequence is that they develop a good sense of everything that is going on and become a source of social intelligence. And the people they talk to feel good as well; how often is someone genuinely interested in what you are doing? It is a flattering experience.

Perhaps the biggest effect of charismatic connectors is not just within teams, but between teams. During Tanzeem Choudhury's PhD thesis research with me, we found that people who characteristically drove conversations—i.e., people who were always curious and asking questions—were the connectors in their organization.[16] They were the ones who moved ideas across group boundaries and kept everyone in the loop. Therefore, these socially intelligent charismatic connectors are key to making organizations successful.

People can teach themselves to be charismatic connectors— they are made, not born. The trick is to do what creative people do: they pay attention to any new idea that comes along, and when something is interesting, they bounce it off other people and see what their thoughts are; they also try to expand their social networks to include many different types of people, so they get as many different types of ideas as possible. They use the coffee pot or water cooler to talk to the janitor, the sales guy, and the head of another department. They ask what's new, what is bugging them, and what they are doing about it, and trade their ideas for ones they've picked up from other people. Not only is it fun to be an idea collector, but people will appreciate help in being kept plugged in.

The bottom line: We have seen how the visualization of interaction patterns can help employees and managers shape idea flow, and consequently improve the productivity and creative output of the organization. By making group members more aware of their patterns of communication within and between groups, we are improving their social intelligence, and this leads to greater productivity and greater creative output.

In the last section, we have also seen that employees and leaders can use their personal patterns of interaction both to directly alter idea flow and to inspire others to develop good habits. Thinking about your job as improving idea flow, getting everyone to talk to each other, and connecting between groups, can be very effective in improving performance.

In the next chapter I will explore how to accomplish these same goals more systematically by using social network incentives. I will show how social network incentives can be used to grow organizations extremely quickly and help them meet the challenges of a changing environment.

· 7 ·

Organizational Change

SOCIAL NETWORK INCENTIVES CAN BE USED TO CREATE INSTANT ORGANIZATIONS AND GUIDE THEM THROUGH DISRUPTIVE CHANGE

Because the social sciences, including economics, have had to work with such impoverished data, it has been difficult for scientists to understand the process of change. Previously, the difficulties involved in collecting sufficient quantities of continuous, fine-grain data have meant that social science analyses were often confined to examining the preconditions of change or large, slow phenomena, such as demographic shifts or long-term health outcomes. The field of economics, for instance, has historically been dominated by analyses of equilibrium states, where everything is in balance (unlike most of the human world).

With the coming of digital media and other big data, all this has changed. We can now watch human organizations evolve on a microsecond-by-microsecond basis and examine all of the interactions among millions of people. When we observe the fine-grain patterns of interaction within an organization, we find mathematical regularities that allow us to reliably tailor the organization's performance and predict how it will react to new circumstances.

In Chapter 2 I highlighted the process of exploration needed to discover ideas and information and showed that social network incentives can be used to ensure that people's exploration was adequately diverse. In Chapter 4 I illustrated the process of engagement used to transform these ideas into behavioral norms by demonstrating that social network incentives can also be used to enforce cooperative behavior. In both cases, the incentives focused on social interactions rather than individual action to drive the community from a failing state into a healthy one.

Now let us examine the Red Balloon Challenge, a case in which my research team and I were able to use social network incentives to build a worldwide organization and accomplish a difficult task in only a few hours, beating hundreds of competing teams to win the prize money. The strategy we took to accomplish this feat was so novel and effective that our approach was published in the journal *Science*,[1] and later expanded upon in the *Proceedings of the National Academy of Science*.[2]

The Red Balloon Challenge was sponsored by the Defense Advanced Research Projects Agency (DARPA) in order to commemorate the fortieth anniversary of the birth of the Internet. The purpose was to discover the best strategies for how the Internet and social networking can be used to solve time-critical search problems. Examples include: search-and-rescue operations in the aftermath of natural disasters; hunting down outlaws on the run;

reacting to health threats that need instant attention; or rallying supporters to vote in a political campaign. The challenge also highlighted the difficulties of dynamic organization building that are typical of large-scale, custom projects such as feature filmmaking or large-scale building projects, but on an incredibly accelerated time scale.

In these sorts of time-critical social mobilization problems it is often not practical, or even possible, to create sufficient mobilization through mass media, due to factors such as the high cost of reaching everyone or infrastructure damage after a disaster. In such cases one has to resort to distributed modes of communication for information diffusion. For example, in the aftermath of Hurricane Katrina, amateur radio volunteers helped relay 911 traffic for emergency dispatch services in areas that experienced severe communication infrastructure damage.

In the Red Balloon Challenge, teams had to find ten red weather balloons deployed at undisclosed locations across the continental United States. The first team to correctly identify the locations of all ten would win a forty-thousand-dollar prize. According to DARPA, a senior analyst at the National Geospatial Intelligence Agency characterized the problem as "impossible by conventional intelligence-gathering methods."[3]

My research group learned about the challenge only a few days before the balloons were deployed, although DARPA had been publicizing it for almost a month. At that point almost four thousand teams had signed up. Despite the stiff competition, we figured that we had a chance of winning, because this was the sort of thing we were expert at, and we quickly formed a team consisting of Riley Crane, Galen Pickard, Wei Pan, and Manuel Cebrian, with help from Anmol Madan and Iyad Rahwan.

Unlike every other team in the competition, our strategy was to reward not only the people who correctly reported the balloons to us, but also those who recruited the eventual balloon finders to our team. Should our team win the $40,000 prize money, we would allocate $4,000 in prize money to each balloon. We promised $2,000 per balloon to the first person to send in the correct balloon coordinates. We also promised $1,000 to the person who invited that balloon finder onto the team, $500 to whoever invited the inviter, $250 to whoever invited that person, and so on. Any remaining reward money would be donated to charity.

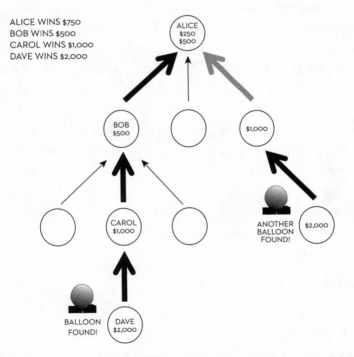

Figure 12. Here, Dave is the first person to report the balloon's location to our team, helping us win the challenge. Once that happens, we send Dave $2,000 for finding the balloon. Carol gets $1,000 for inviting Dave, Bob gets $500 for inviting Carol, and Alice gets $250 for inviting Bob. The remaining $250 is donated to charity.

Our social incentive approach differs from the direct, market-based approach of giving $4,000 per balloon in two key ways. First, a direct reward might actually deter people from spreading the word about our team, as any new person recruited would represent extra competition for the reward. Second, the direct approach eliminates people living outside the continental United States, as there is no possibility of them spotting a balloon.

These two factors played a key role in our success, as illustrated by the fact that recruitment chains were up to fifteen people long, and approximately one of three tweets spreading information about the team originated outside the United States. Distributing the reward money more broadly motivated a much larger number of people (more than five thousand) to join the team, including some from outside of the United States who could be rewarded for simply e-mailing someone who could find a balloon. Perhaps even more impressive is that we estimate that each of our five thousand team members alerted an average of four hundred friends each, for a total of almost two million people helping us to search for the red balloons.

As a result of using this social network incentive strategy, our research team correctly identified the location of all ten balloons in just 8 hours, 52 minutes, and 41 seconds.

Instant Organizations

At first glance, many people think the Red Balloon Challenge is an example of crowdsourcing, similar to Amazon Mechanical Turk, which allows users to "hire" thousands of independent people to do simple, individual tasks. But that is old-fashioned market thinking, and that type of strategy failed for other teams participating in the Red Balloon Challenge.

The point is not just that it's possible to get lots of people to work. Rather, the point is that it's possible to get people to build an organization that does the work. That is why we rewarded people both for finding balloons and for recruiting people to help search. We rewarded people roughly equally for these two tasks, because building the network was just as important as the actual work of searching. We used a standard individual economic reward to motivate people to report balloons to us but a social network reward to get them to recruit more people. Our approach to the Red Balloon Challenge was startling because it showed that we could build an organization of thousands of people, accomplish an extremely difficult task, and finish successfully in only a few hours.

As a comparison, consider Wikipedia, an immense online encyclopedia built by volunteers that is often cited as a great example of crowdsourcing. While it is true that many people contribute content, there was also a core group of dedicated editors who worked for years to organize the content as it was added. These editors were recruited in a fashion similar to the way we recruited for the Red Balloon Challenge, although the network incentives in the Wikipedia case were social rather than monetary. Just as in our FunFit example, the behavior of these editorial recruits was shaped by social network incentives, until they became a deeply engaged working group with standardized, shared ways of doing things.[4]

Imagine building Wikipedia using a true crowdsourcing tool like Mechanical Turk, with which there are no network interactions or social incentives. In that scenario, the workers don't know each other, and perhaps they just get an e-mail telling them what the next task will be. Thousands or millions of workers would be paid to independently manufacture content, then hundreds more would be paid to review the content for quality and completeness, still others would be hired and paid for any necessary edits, and then finally a small central managing board would still probably be needed to set policies and decide how to do all this. It is easy to see that this sort of hierarchical crowdsourcing would be inefficient and expensive, and the result would likely be a real mess.

Nevertheless, during the last century this sort of hierarchical crowdsourcing has been exactly the model of most corporations. Workers sit in cubicles doing independent tasks, and then their outputs are routed to anonymous others for the next stage of processing. Other anonymous workers then use checklists for quality control, and finally, a central management oversees the whole affair. This is why traditional encyclopedias were so expensive to produce, and why today most corporations are still inefficient and slow to change.

The central problem illustrated here is that these sorts of old-fashioned organizations were built using market thinking: incentivize the faceless, identical workers to produce cookie-cutter pieces of work. Because these types of organizational structures incorporate few or even no peer-to-peer network incentives, workers tend not to help each other learn best practices or maintain high levels of performance. And because workers are not engaged with management, neither has the opportunity to learn from the other, and so business processes remain rigid and inefficient. In contrast, in Wikipedia's organization, continual exchanges between contributors and editors led to the development of interaction patterns that evolved to meet the needs of this rapidly growing enterprise. Peer pressure around these habits of interaction produced coordinated activity in a very efficient, effective manner.

Organizations in Stress

The Red Balloon Challenge illustrates an extreme sort of organizational dynamics. All organizations are dynamic in some way, though, and have predictable changes in their network of idea flow as the organization reacts to new events and circumstances.

Consider the process of engagement, which drives the creation of organizational habits and enforces those habits with social pressure. When a group is faced with change, it needs to create and enforce new habits of interaction that will help it adapt to the new circumstances.

A new product, a new computer system, or a reorganization of the company usually means that everyone's job changes. As a result, who everyone has to coordinate with, the details of what they do, and how they divide up the work may change. All of this means that there is an urgent need to create and adopt new habits and therefore a greater need for engagement within the group.

This systematic change in the employee's level of engagement in response to new challenges is exactly what we observe in our research. For instance, when my group measured the interaction patterns within a German bank, it serendipitously happened that the employees were suddenly faced with a much higher workload (see Figure 8).[5] By using the second-by-second sensing abilities of the sociometric badges, we could see that the level of engagement between employees almost immediately shot up, helping them develop new patterns of working that would enable them to handle the increased workload.

We observed another example of engagement in the service of developing organizational habits while monitoring a travel company of about 120 employees while they went through a round of layoffs.[6] Immediately afterward, we could see in our fine-grained, second-by-second data that the level of engagement immediately skyrocketed, as the remaining employees began to adapt to the new situation by generating new patterns of interaction. Interestingly, it was the employees with the greatest engagement before the layoffs who had the easiest time adapting to the new patterns of interaction.

The increases in engagement we observed might be considered just people providing social support. In fact, they do much more than that, because as we have seen in the last two chapters, changes in engagement also cause changes in the level of productivity. High-stress situations lead to greater engagement levels almost immediately, as people begin to talk to each other in order to figure out what to do, and then begin the task of forging new patterns of interaction that are better adapted to the situation. Later, changes in the network of interactions act like social network incentives, as the desire to reduce stress drives the development of new patterns of interaction.

Trust

Trust is developed by stable, frequent interactions with others, so social networks pioneer Barry Wellman suggested that a rough-and-ready measure of the strength of social ties might be the frequency of social contact.[7] In Chapter 4, we saw that Wellman was absolutely right: The frequency of direct interactions in our Friends and Family Study accurately predicted the level of trust between two people. And increased trust enables a greater flow of ideas and thus leads to greater productivity.

The strength of social ties was a critical variable in the engagement experiments in Chapter 4 as well as in the Red Balloon

Challenge. In both cases, cooperation was most effective when it leveraged preexisting personal social ties, and the more active the social tie, the greater the level of cooperation. For the Red Balloon Challenge, invitations along the most active social ties were more than twice as effective for recruiting participants than the average invitation. In the FunFit experiment, the social influence exerted by buddies with active social ties was more than twice as effective as that from buddies who were only acquaintances.

I believe that it was the investment in social ties in our Red Balloon Challenge incentive structure—sometimes called building social capital—that mattered most, rather than individual financial incentives. For the average participant, the expected financial reward for participation was near zero. There were literally millions of people aware of the challenge and on the lookout for red balloons, and thousands of teams signed up to compete. So the odds that a participant or one of their recruits would be first to report a balloon to the winning team were a million to one, and yet thousands of people signed up their friends to help us search.

Our Red Balloon Challenge follow-up interviews suggested that people signed up their friends as a favor to the friends. That is, recruiting a friend was like sharing a free lottery ticket. You don't necessarily expect to win, but sharing the ticket strengthens the social ties with your friend. By sharing, you make it more likely that they will share with you or help you out on another occasion; you are building trust and social capital.

Building strong ties with people is good for idea flow, but strong ties also can be used to exert social pressure. In the engagement experiments of Chapter 4, people were rewarded when their buddy did well. The pairs of people who had the most invested in the relationship, that is, those who interacted and cooperated the most, were the ones who could exert the most social pressure on each other. Or to put it another way, we really don't want to make our office mates or our moms mad, and that is why we compromise with them when they want us to change our habits. As the experiments in Chapter 4 showed, strong social ties create the conditions in which peer pressure is the most effective mechanism for promoting cooperation.

This connection between engagement, trust, and people's ability to act cooperatively is perhaps the main point of Robert Putnam's classic book *Bowling Alone*, which highlights the relationship between civic engagement and the health of society.[8] We are traders in ideas, goods, favors, and information and not simply the competitors that traditional market thinking would make us. In each area of our lives, we develop a network of trusted relationships and favor those ties over others. Exchanges within this network of trusted social ties facilitate idea flow, creating an inclusive, vigorous culture, and are responsible for the collective intelligence of our society. In Chapter 5 I showed how the flow of ideas is directly responsible for increases in productivity and creative output within groups and companies. In Chapter 9 I will show that the same is true of entire cities.

Understanding ourselves this way could have a dramatic effect on the character of our society. Because idea flow creates culture, supports productivity, and enables creativity we should place greater value on professions that enhance idea flow: teachers, nurses, ministers, and policemen, along with doctors and lawyers who work for charities, as public defenders, or for inner-city hospitals. Better rewards for work that reinforces our social fabric would allow us to find a better, more sustainable blend between individual ambitions and the health of society.

Next Steps

In the last three chapters we have seen how idea flow affects the collective intelligence of work groups and organizations, and how idea flow can be improved by use of visualization. Finally, in this chapter we have discussed how social network incentives can be used to grow organizations and help them through change. In the next section of this book I will show how to apply these same social physics concepts to cities. My goal is to imagine what a data-driven city might look like and how we can use big data and social physics to create more productive and creative cities. And then, in the last section, I will discuss what changes need to be made to privacy, management, and government in order to create a brighter, safer future.

SOCIAL SIGNALS

•

In my book on interpersonal communication, *Honest Signals*, I showed that patterns of interaction independent of content are an accurate measure of idea flow and decision making.[9] As reported in *Nature*, this happens because the pattern of interaction—who interrupts whom, how often people talk, and to whom—are social signals of dominance, idea flow, agreement, and engagement.[10]

Therefore, usually we can completely ignore the content of discussions and use only the visible social signals to predict the outcome of a negotiation or a sales pitch, the quality of group decision making, and the roles people assume within a group.

How do these social signals interact with language in modern humans? Evolution rarely discards successful working parts. It generally either builds additional structures while retaining the old capabilities or subsumes old structures as elements of the new. When our language capabilities began to evolve, our existing signaling mechanisms were incorporated into the new design. As a consequence, our ancient social signals still shape our modern patterns of conversation.

In several studies of small groups solving problems, we instrumented each person to measure both their social signaling and their pattern of interaction. We found that each of the different social roles that psychologists identify, i.e., protagonist, supporter, attacker, or neutral, uses different social signaling and, as a consequence, different patterns of speaking length, interruption of others, frequency of speaking, etc.[11]

The same is true of the information content: Someone contributing a new idea speaks differently than someone who is orienting the group to return to a previous idea or someone who is neutral. As a result, each person's pattern of interaction can be used to identify their functional role—follower, orienteer, giver, seeker, and so on—without listening to the words.

Similarly, just as social signals determine the dominance network in ape colonies, the patterns of conversations in modern humans determine people's places in their social networks. In particular, the social structure can be figured out from the pattern of who controls conversation, that is, who initiates conversation, who interrupts, etc.[12]

For instance, when we measured the conversational patterns of twenty-three members of our laboratory over a period of two weeks, we found that influence over the conversation pattern predicted the subjects' position in their social network nearly perfectly.[13] Together, these and other experiments strongly suggest that influence over the pattern of conversation is an accurate measure of individuals' influence in their surrounding social networks.

We are all familiar with many of these social signals, but others are more difficult for us to perceive consciously. A familiar example is mood contagion.[14] If one member of a group is happy and bubbly, others will tend to become more positive and excited. Moreover, this signaling-induced effect on mood serves to lower perceptions of risk within groups and to increase bonding.

Similarly, people tend to mimic each other automatically and unconsciously.[15] Despite being mostly unconscious, this mimicking behavior has an important effect on its participants: It increases how much they empathize with and trust each other. Not surprisingly, negotiations with lots of mimicry tend to be more successful, no matter which party starts copying the other's gestures first.

Each of these signals has roots in the biology of our nervous system. Mimicry is believed to be related to cortical mirror neurons, parts of a distributed brain structure that seems to be unique to primates and is especially prominent in humans. For example, mirror neurons react to other people's actions and provide a direct feedback channel between people. One result of this is the surprising ability of human newborns to mimic their parents' facial movements despite their general lack of coordination.

Similarly, our activity level is related to the state of our autonomic nervous system, an extremely old neural structure. Whenever we need to react more vigorously—say in fight-or-flight situations or when sexually aroused—this system increases our activity level. On the other hand, we tend to be listless and less reactive when our autonomic nervous system is blunted, as during clinical depression. The relationship between autonomic nervous system function and activity level is tight enough that we have used it to accurately estimate the severity of depression.

Indeed, these signaling patterns are so clear that they are now used commercially to screen for mental health conditions such as depression and to monitor patient engagement during treatment. For more detail see http://cogitocorp.com, an MIT spin-off company that I cofounded.

PART THREE

Data-Driven Cities

•

· 8 ·

Sensing Cities

HOW MOBILE SENSING IS CREATING A NERVOUS SYSTEM FOR CITIES, ENABLING THEM TO BECOME MORE HEALTHY, SAFE, AND EFFICIENT

Sustaining a healthy, safe, and efficient society is a scientific and engineering challenge that goes back to the 1800s, when the Industrial Revolution spurred rapid urban growth and created huge social and environmental problems. The remedy then was to build centralized networks that delivered clean water and safe food, enabled commerce, removed waste, provided energy, facilitated transportation, and offered access to centralized health care, police, and education services.

But these century-old solutions are increasingly obsolete. We have cities jammed with traffic, worldwide outbreaks of diseases that are seemingly unstoppable, and political institutions that are deadlocked and unable to act. In addition, we face the challenges of global warming; uncertain energy, water, and food supplies; and a rising population that will require building one thousand new cities of a million people each in order just to stay even.

But it doesn't have to be this way. We can have cities that are energy efficient, have secure food and water supplies, and much better government. To reach these goals, however, we need to radically rethink our approaches. Rather than static systems that are separated by function—water, food, waste, transport, education, energy, and so on—we must consider them as dynamic and holistic systems. We need networked, self-regulating systems that are driven by the needs and preferences of the citizens instead of ones focused only on access and distribution.

To ensure a sustainable future society, we must use our new technologies to create a "nervous system" that maintains the stability of government, energy, and public health systems around the globe. Our current digital feedback technologies are already capable of creating the level of dynamic responsiveness that our larger, more complicated modern society requires. We must use these technologies to reinvent societies' systems within a control framework: one that first senses the situation; then combines these observations with models of demand and dynamic reaction; and, finally, uses the resulting predictions to tune the systems to match the demands being made of them.[1]

Right now, the most important generator of city data is a familiar tool: the ubiquitous mobile phone. These devices are, in effect, personal sensing devices that are becoming more powerful and more sophisticated with each product iteration. In addition to deriving information on user locations and call patterns, we can map social networks, and even gauge people's moods by analyzing the digital chatter that has become so pervasive. Consumers are also beginning to make purchases simply by scanning items with their phones, thereby adding financial and product choice information to the digital biographies sketched by mobile phone traffic. Moreover, as smartphones continue to morph into personal information hubs that have greater computing capacity, they will reflect ever more information about human behavior.

Together, wireless devices and networks constitute the eyes and ears of this evolving digital nervous system. Furthermore, its evolution will continue at a quickening pace because of the exponential progress in computing and interaction technologies, as well as of basic economic forces. Networks will become faster, devices will have more sensors, and techniques for modeling human behavior will become more accurate and detailed.

Many of the sensing and control elements required to build a digital nervous system are already in place. What is missing, though, are two critical items: The first is social physics, specifically dynamic models of demand and reaction that will make the system function correctly, and the second is a New Deal on Data, an architecture and legal policy that guarantees privacy, stability, and efficient government. I will discuss the social physics side of the digital nervous system in this and the next chapters, but I will wait until the final chapters to address privacy, stability, and efficiency.

Behavior Demographics

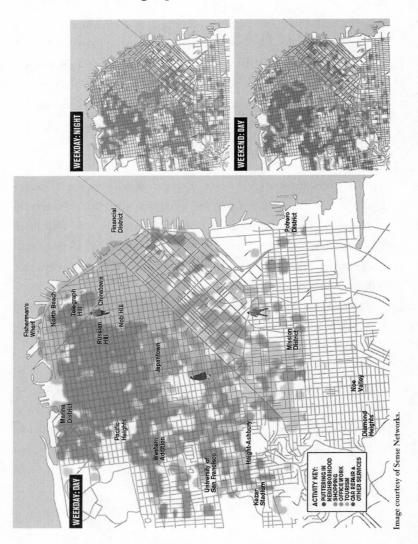

Figure 13: Reality mining of GPS data from mobile phones. Human activity patterns within a city, gray-level coded by common patterns of activity. The patterns of activity reveal distinct rhythms that change in predictable ways.

Today commercial operations and government services all rely on demographic data to guide them. Which neighborhoods are residential? Which are industrial? How many people work or live there? How wealthy are they? Unfortunately, these data are currently expensive to collect. For example, in the United States the government census is done only once every ten years and it can quickly become out-of-date. And in many parts of the world, these demographic data simply do not exist at all. The proliferation of mobile phones makes it possible to leap beyond demographics to directly measure human behavior. Using data gathered from the digital bread crumbs that people leave behind, we can more readily answer questions such as: Where do people eat, work, and play? What routes do they travel? Who do they interact with?

Figure 13 shows activity patterns within San Francisco based on GPS data from mobile phones that were collected by San Francisco's Exploratorium, its science museum. The patterns are gray-level coded by common types of activity (such as shopping, chores, and tourism) among restaurants, entertainment venues, service shops, businesses, and so on. These activities have rhythms that are predictable across days and weeks. These data, created by an MIT spin-off company, Sense Networks (which I cofounded), allow us to analyze the movement and purchasing behaviors of tens of millions of people in real time.[2]

What Figure 13 does not show is that the population is made up of distinct subgroups, sometimes called tribes. Members of each tribe choose to go to the same places, eat similar foods, and enjoy the same entertainments. These choices place them into a behavior demographic, because the behaviors of the tribe members selectively reveal their underlying preferences. Moreover, because people in these tribes spend time around each other, the process of social learning gains traction and drives the development of behavior norms within the given tribe. These may have nothing to do with their conscious preferences, and tribe members may not even be aware of them at all. But nevertheless, people within the same behavior demographic have similar food habits, similar clothes, similar financial habits, and similar attitudes toward authority, and as a consequence, they have similar health outcomes and similar career trajectories.

In my experience, these behavior demographics typically provide predictions of consumer preferences, financial risks, and political views that are more than four times as accurate as standard geographic demographics based on zip codes. They also accurately predict people's risks for diseases of behavior, such as diabetes or alcoholism. As in the Friends and Family and Social Evolution studies discussed in Chapter 3, the process of social learning and the development of social norms within cities is driven by the observation of peer behavior, that is, by people trying to fit in with their chosen peer groups.

In addition to knowing about people's preferences, we can also better understand the rhythm of their daily habits. Most people have many constraints on how they spend their time; there are times for work, times for sleep, and times for play. Mealtimes, coffee breaks, and time with friends follow daily and weekly patterns. But what are the overall patterns of our lives? The timing of our to-ing and fro-ing sets the rhythms of the city, as shown in Figure 13, and determines peak demands on transportation, energy, entertainment, and sustenance.[3]

For most people, the primary pattern is the workday, that is, going to work and coming home, usually along the same path day after day. The second most pronounced pattern is the weekend and days off, often with the characteristic behavior of sleeping in and spending that night out in a location besides home or work. Perhaps surprisingly, the places we go and things we do during our free time are almost as regular as our work patterns. The third pattern, however, is a wild card: days spent exploring, usually a shopping trip or an outing. This last is distinguished by its *lack* of structure. Together these three patterns typically account for 90 percent or more of our behavior.

By combining these habits in time with the behavior demographics discussed earlier, we can create a society that is much better managed. Knowing the typical behavior patterns within a city can allow us to better plan city transportation, services, and growth. Specifically, continuous streams of data about human behavior allow us to accurately forecast changes in traffic, electric power use, and even street crime and the spread of the flu. As we will see in the next few sections, these data-driven forecasts allow us to prepare for peaks in demand and manage them better. It also means that we can react better to emergencies or disasters, because we can know who is likely to be where and when. The ability to know where and when the people who are at risk of diabetes eat, or where the people who have trouble handling money shop also has great potential to improve public health and public education.

In the next few sections I will describe how this new nervous system can change our lives, primarily using examples from the areas of transportation and health. Examples of transforming government, creating learning environments, and enhancing the creativity of our culture will be presented in later chapters.

Transportation

A familiar example utilizing the digital bread crumbs people leave behind is the use of GPS data collected from drivers' cell phones to provide minute-by-minute updates on traffic flow. This enables more accurate detection of traffic congestion patterns and driving time estimates; simple versions of this capability are already built into car navigation systems worldwide. It is easy to imagine improving on today's systems by marrying these data with information from each individual driver's daily calendar in order to generate personal transportation schedules that avoid traffic snarls. Similar schedules could be generated for commercial traffic by moving delivery and commuting traffic to different times and routes, thereby increasing the efficiency of distribution networks.

These sorts of applications, however, only scratch the surface of what might be possible. When looking at data from cell phone systems in the cars themselves, e.g., systems such as GM's OnStar, we see that we can do a good job of predicting when someone is in danger of having an accident. A simple example consists of basically crowdsourcing dangerous conditions: If other cars have just recently gone down the road you are driving on and had emergency braking events, then you are at significant risk of an accident. If you are traveling faster than the other cars were, then you are in real danger. Warnings based on this sort of big data could be used to reduce accident rates dramatically.[4]

It also is possible to dramatically improve the flows that keep our cities alive—the stuff moving through cities by trucks, trains, and pipes—by combining habit and preference data with those on weather, among other information. Being able to anticipate the rhythms of a city helps companies to prepare for the peaks and valleys in demand, as well as to streamline distribution networks. The typical city bus system gets only around one mile per gallon per person in fuel efficiency, except at rush hour, but we have to keep those huge buses on the street, because unexpected crowds of passengers occasionally materialize. Urban planning could be improved as well, since knowing when and where citizens are likely to go means that we can plan city growth to minimize traffic and energy use, while at the same time improving convenience.[5]

Perhaps the most interesting idea is to use transportation networks to increase the productivity and creative output of cities: We can use data about people's habits to structure public transportation to promote more exploration within a city. It has long been observed that physically isolated neighborhoods have worse social outcomes.[6] This is linked to the concept developed in Chapter 2, in which exploration among groups improves both productivity and creative output. On the scale of cities, it suggests that the number of different neighborhoods that can be visited conveniently sets the pace of exploration and thus the pace of innovation and productivity growth. Designing a city for fast, flat-rate transportation that promotes both village-style neighborhoods and central big business and cultural areas may be the simplest and cheapest way to both improve poor neighborhoods and increase overall productivity, as I will discuss in the next chapter.

Health and Disease

In the area of public health, a current, familiar example of using big data to engineer a better society is Google Flu, which predicts outbreaks by counting the number of Internet searches using the word "flu" that occur in each state or region. Regions with a strong increase in the number of these online searches are likely to be experiencing increases in the number of cases. Techniques such as this are important in assisting the Centers for Disease Control to detect new strains of flu, as well as for predicting the amount of medicine that will be required and for helping hospitals, cities, and companies anticipate the number of sick that they will have.

Again, this is just scratching the surface of what a digital nervous system can contribute to public health. Before now, doctors never had ways of quantitatively measuring behavioral changes that occur when a person is becoming ill. Thus, most research on infection spread typically assumes that movement and interaction patterns change very little during an infection; that is, ill people still mostly continue about their daily patterns of behavior.[7]

But mobile phone data show that this is not true: People reliably change their behaviors when they become ill. PhD students Anmol Madan and Wen Dong and I have discovered that people's behaviors change in regular, predictable ways when they are becoming ill, and that we can measure these behavioral changes using the sensors in mobile phones.

With reported sore-throat and cough symptoms we found that people's normal patterns of socialization were disrupted, and they began to interact with more and different people (good for the virus but bad for humans). With common colds we found that total interactions, and nighttime interactions, increased: People seemed to be calling their friends after work hours.[8]

Later in the disease cycle, when there were fever and other influenza symptoms, we found that people limited their movements dramatically (good for the other humans). People who reported feeling stressed, sad, lonely, or depressed became socially isolated on symptomatic days. All these examples illustrate the tremendous potential of mobile phones to monitor the health status of an individual in almost real time.

Because behavior changes associated with problems such as respiratory symptoms, fever, influenza, stress, and depression are similar for everyone, but each problem is different from all the others, it is possible to actually classify people's overall health state from their behavior alone. For example, an app on a phone could quietly look for uncharacteristic variations in behavior, and then figure out if an illness is developing. Such proactive health care could be critical for conditions with high risk of patient underreporting (e.g., of declining mental health or aging-related problems). This is the idea underpinning another of my group's spin-off companies, Ginger.io, that I helped cofound with my student Anmol Madan, based on his PhD research.[9]

Going even further, by crowdsourcing this behavior information across a population, and then combining that information with data about where and when people went during the previous days, the infection risk of an entire area can be figured out, as illustrated by the map in Figure 14.[10] This shows where people are most and least likely to become infected with the flu at a particular day and hour.

The ability to track diseases such as the flu at the level of individuals would give us real protection against pandemics, because we could take steps to reach infected people before they spread the disease further. The way real-time flu tracking would work is by combining information from two sources: 1) data about changes in individuals' behavior patterns, because when people are becoming ill, we can measure predictable changes in these patterns; and 2) location data, because physical interactions with others are the primary mechanisms for propagation of airborne contagious diseases.

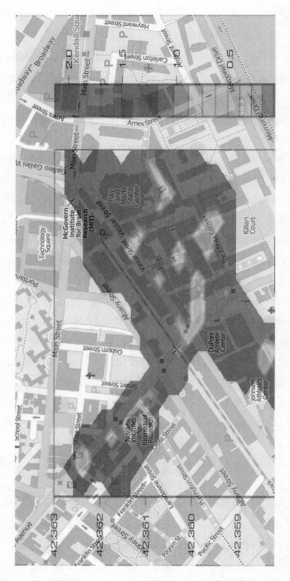

Figure 14: A map of people's interactions at each location and how likely they are to catch the flu. The dark areas are where we have data; the lighter areas within the dark are where catching the flu is more likely.

Specifically, we can use the knowledge of how people's behaviors change when they are becoming ill and measure them using the sensors in mobile phones to estimate the likelihood that each individual is becoming ill. Wen Dong has shown that when we combine these individual likelihoods by mathematically modeling the spreading process we can create maps such as the one shown in Figure 14. Because this map shows the level of infection danger in each location, it can be used to avoid the places flu exposure is most likely.[11]

The need to be able to track disease at the level of the individual and in real time is becoming increasingly urgent. As the world becomes ever more interconnected through the movement of people and goods, the potential for global pandemics of infectious diseases rises as well. In recent years, outbreaks of Severe Acute Respiratory Syndrome (SARS) and other serious infectious diseases have spread quickly between widely separated but socially linked communities. As a result, the danger of pandemics, be it SARS or H1N1 or some other infectious disease, has increased dramatically.

The capacity to see the spread of infectious disease on a person-by-person, minute-by-minute basis gives us the ability to take genuinely effective preventive action. In fact, some experts on infectious disease believe that this is one of the only hopes we have for avoiding hundreds of millions of deaths from the pandemics that are certain to come.[12]

Social Network Interventions

Within our vision for data-driven cities, however, lies a persistent challenge: How can we build systems that people will actually use? Unless a system is compatible with human nature, people will either ignore or misuse it. Building a human-centered city that is also driven by data requires integrating the lessons of social physics into every aspect of it.

Current city systems designs typically rely on financial incentives: Road tolls are greater in the centers of towns, taxes are higher than in the suburbs, and so forth. Unfortunately, experience shows that this approach rarely works very well, especially in tragedy of the commons situations.

Moreover, using financial incentives privileges the rich. As an example, consider congestion pricing as a method of managing traffic. By charging people more to drive in certain places, we allow rich people to go where they want and keep out the poor. This is particularly worrisome because exploration results in innovation, so by reducing the amount of exploration that the poor can achieve, we are also reducing their community's capacity for development and social improvement.

In contrast, a social physics approach to building social norms relies on influencing the social network. There are three types of interventions that are naturally suggested by the social physics perspective.

Social mobilization: As used in the Red Balloon Challenge (see Chapter 7), social mobilization is critical for tasks such as searching for missing kids or escaped criminals, and for finding critical supplies after a disaster such as an earthquake or tornado. Recall that in our solution for the Red Balloon Challenge, social network incentives enabled us to recruit many people to solve a problem in a very short period of time.

I imagine that the primary use of this type of incentive will be to create new organizations rather than to solve short-term crises. Already I see this type of incentive being used in political campaigns to attract grassroots workers and by start-ups to recruit new employees.

Tuning the social network: A second type of intervention involves tuning the network to provide sufficient idea diversity. In Chapter 2 I showed that people made much better decisions when they could see those of a wide range of other people, and their outcomes. The exception to this wisdom of the crowd phenomenon was when the social network was so dense that it formed a sort of echo chamber, so that the same ideas circulated around and around.

To solve the problems of both insufficient diversity and echo chambers, we were able to tune the flows of ideas between people by providing small incentives, or nudges, to individuals. These caused isolated people to engage more with others, and caused people who were too interconnected to engage less and to explore outside their current contacts.

We are now beginning to apply this tuning idea to other social networks. For instance, imagine tuning the advice networks some companies have deployed to collect the wisdom of the crowd from employees. The company's goal is to run more smoothly, and people are supposed to record how they tried to solve problems and then what happened, rather like product reviews found on many Web sites. Some companies even offer an economic incentive for posting advice: If a particularly useful idea is posted, then the poster is actually paid for sharing.

In addition to providing feedback, however, the pattern of links between ideas, reactions, and further suggestions provides an accurate picture of the networks along which ideas are propagated. This then facilitates the measurement of idea flow, enabling us to measure the pattern of ideas and reactions and to see whether or not a sufficiently diverse set is being considered and to determine if effective social learning is occurring. As a result, we can tell people when the set of ideas they have looked at is sufficiently diverse to help them make a reliable, good decision.

We can also create diversity ratings of news blogs and similar civic media, so that no one interest group can drown out everyone else. These sorts of tuning interventions are important for addressing some of the ills of our new, hyperconnected world. Today fads and panics seem to ripple continuously through our societies, causing overreactions and stress and distracting us from the slower, more patient work of building a better world. Perhaps by tuning our news networks to reduce recirculation of rumors and spin we could better focus on making real progress.

L everaging social engagement: This third type of network intervention is helpful in addressing tragedy of the commons situations by using social network incentives to increase engagement around the problems within local communities. In Chapter 4, we saw that giving rewards for improvements in other peoples' behavior produced social pressure for cooperation and that this social pressure reliably caused larger behavior changes than giving people rewards to change their own behavior.

The same ideas can be applied on larger scales. In Chapter 4 we also saw how a Facebook "get out the vote" campaign in 2010 targeted 61 million people. The direct effect of this campaign was not terribly large, but by letting people share "I Voted" postings with their friends, they produced social pressure in face-to-face social networks, and that substantially increased the number of people who voted. Another example from Chapter 4 was when my research group and our colleagues at ETH deployed a social network as part of the electric utilities' Web pages and encouraged people to form local buddy groups. This network used social incentives rather than standard economic ones: When people saved energy, gift points were given to their buddies. The social pressure this created caused electricity consumption to drop almost 17 percent—twice the improvement seen in earlier energy conservation campaigns.[13]

From a Digital Nervous System to a Data-Driven Society

Today we have a digital nervous system of sensors and communication already in place, ready to transform our cities into data-driven, dynamic, responsive organisms.[14] Great leaps in health care, transportation, energy, and safety are all possible.[15] In the Data for Development project that I will describe in Chapter 11, we will see how even with only low-resolution, anonymous, and aggregate data, researchers were easily able to find transportation improvements of more than 10 percent, health improvements of more than 20 percent, and make important contributions to the problem of reducing ethnic violence. The main barriers to achieving these goals are privacy concerns and the fact that we don't yet have any consensus around the trade-offs between personal and social values.

We cannot ignore the public goods that such a nervous system could provide. Hundreds of millions of people could die in the next flu pandemic, and it appears that we now have the means to contain such disasters. Similarly, we are able not only to reduce energy use in cities dramatically, but as we will see in the next chapter, we can even shape cities and communities to both reduce crime and at the same time promote greater productivity and creative output. The key, as you might expect, is using social physics to shape idea flow.

· 9 ·

City Science

HOW SOCIAL PHYSICS AND BIG DATA ARE REVOLUTIONIZING OUR UNDERSTANDING OF CITIES AND DEVELOPMENT

Thomas Jefferson famously referred to eighteenth-century cities as "toilets of all the depravities of human nature." But since Jefferson's day, the cities of the world have grown a hundredfold, and growth continues unabated. A larger percentage of people now live in cities than at any other point in human history.[1] Why do people persist in moving to cities despite soaring living costs and elevated levels of crime, pollution, and infectious disease?[2] It may be that Adam Smith called it correctly: Urban centers are exceptional not only for depravity but also for innovation.[3]

Despite more than a century of intense study of cities, we still lack a compelling model for why urban areas tend to promote innovation. And innovate, they do. Urban areas use resources more efficiently and produce more patents and inventions with fewer roads and services per capita than rural areas.[4] What is it about having more people living together that leads to the more efficient creation of ideas and increased productivity? Some people point to the role of technology diffusion in creating intellectual capital,[5] while others argue about the role of hierarchical social structures and specialization.[6]

The Social Physics of Cities

As I have discussed in earlier chapters of this book, social network interactions and idea flow are major drivers of creative output and productivity in groups and companies. These social physics concepts are almost uniquely scalable within the social sciences, and as I will show in this chapter, they extend beyond small group and corporate dynamics to work at the scale of cities, promoting greater productivity and creativity throughout these much larger social networks. Cities are idea machines in the same way that companies are idea machines.

Together with students and colleagues Wei Pan, Gourab Ghoshal, Coco Krumme, and Manuel Cebrian, I have developed a mathematical model for how social ties drive idea flow within cities based on the number of people within face-to-face meeting distance. As we described in *Nature Communications*, this model gives us a simple, bottom-up, robust model that quantitatively predicts GDP and creative output.[7] We have also been able to show that idea flow along social ties accurately reproduces urban features, such as the rates of HIV/AIDS infections, telephone communication patterns, crime and patenting rates, and more. It also gives us insights about how to engineer cities to be both more creative and productive while at the same time minimizing crime and other negatives.

It is important to note that this social physics view of cities is different from the classic models of class and specialization in that it focuses on idea flows rather than static divisions in society. In this way, social physics is similar to models that explain the manufacturing efficiency of cities in terms of the proximity of factories and the costs of transporting goods.[8] The difference, however, is that social physics conceptualizes cities and companies as *idea* factories, so the focus is on the flow of ideas rather than the flow of goods.

In taking this perspective, social physics is part of a long line of thought in sociology, geography, and economics that explores the relationship between population density and innovation, as well as diffusion along social ties and creativity.[9] The new and important contribution that social physics brings is the integration of these ideas into a single mathematical model that can be tested against dense, continuous behavior data and available economic and social outcome data. Social tie density and idea flow offer simple, generative links among human interaction patterns and mobility patterns and the characteristics of urban economies without the need to appeal to hierarchy, specialization, or similar social constructs. As the remainder of this chapter will explain, what really matters is the flow of ideas, not classes or markets.

Social Ties in Cities

The pattern of social ties within a city is well described by the notion that the chance of a relationship between two people is determined by the number of "intervening opportunities." At its core, this is just the simple idea that the chance of starting a friendship with a particular stranger in a crowd is smaller if there are a lot of other potential friends in the crowd. For example, Liben-Nowell and colleagues studied people who were members of a diary Web site and mapped how far away their friends and acquaintances lived.[10] They found that for the majority of friends, the likelihood of two people forming a social tie drops off smoothly in proportion to the number of people who spend time at locations between them.[11] A similar relationship has been found in the location-based social network app Gowalla, which records where both individuals and their friends "check in." These data allow researchers to see how close to each other friends live, as well as how often friends go to the same places.[12] The result of this research is a simple mathematical equation that describes how people tend to have lots of social ties to people who live nearby and increasingly fewer ties with people who are farther and farther away.[13]

This mathematical relationship concerning social ties, however, has other, more interesting applications. For instance, it seems clear that the spread of diseases such as HIV/AIDS[14] depends on the distribution of social ties, and so (in a very different way) does the pattern of telephone calls.[15] Can these two very different phenomena, i.e., the pattern of phone calls as a function of county population and the frequency of HIV/AIDS cases as a function of the population density per square mile, both be predicted by the same mathematical relationship between distance and number of social ties as measured by analyzing Web sites and social networks?

Figure 15 shows that our model of social ties accurately describes how both of these social patterns change with increasing population density. These figures show that the same simple mathematical model for how social ties are linked to distance produces consistent predictions across physical, telephone, Web, and social network interactions. A quantitative, predictive model that

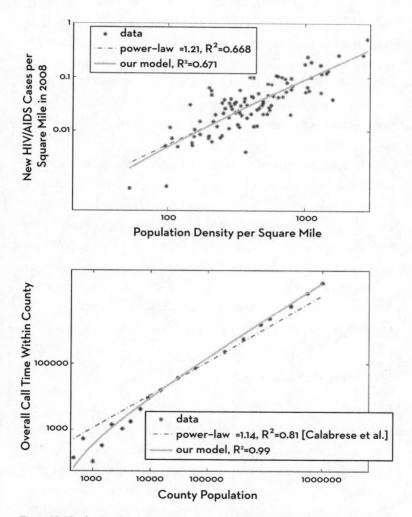

Figure 15. The density of social ties accurately predicts both telephone calling patterns and the rate of HIV/AIDS cases.

cuts across so many phenomena and across such a wide range of scales is quite rare in any field of science and almost unheard of in the social sciences.

All of these social tie patterns in cities have their counterparts in smaller group structures. Just as we have seen in our work in companies (Chapters 5 and 6), our closer social ties support engagement, because those people are more likely to talk to each other and provide the reinforcement that transforms ideas into behaviors. Likewise, our distant social ties serve the role of exploration, because we meet new people in new contexts and harvest new ideas from them.

For companies, however, there is usually a sharp boundary between the work group and "others." In the rest of our lives there is usually no sharp boundary between exploration and engagement in our overall patterns of interaction with other people. That is, when we look at all of our interactions we see that people have many social roles (e.g., mother, coworker, citizen, jazz enthusiast, etc.), and each role engages a different set of people, so that the functions of engagement and exploration are combined across all of a person's social networks.

Exploring the City

In previous chapters I have described studies conducted by my research group that utilized data from big data sources such as mobile phones, social networks, and sociometric badges. Another big data lens for examining human behavior is credit card data. Through an agreement with a major U.S. financial institution, Coco Krumme and I were able to analyze credit card use statistics from almost half of the working adults in the United States during

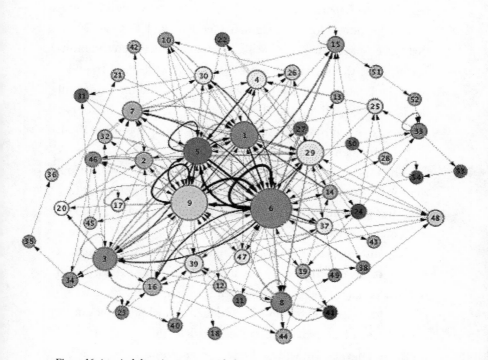

Figure 16: A typical shopping pattern, with the size of each circle indicating the frequency of places visited. Arrows show the frequency of transitions between places. The frequency of visitation drops off smoothly and decreases with distance. The stores, restaurants, and entertainment places that you visit most are also likely to be visited by your friends, and so are unlikely to contribute new ideas to you or your social network. The greatest chance of an experience that is new to everyone in your social network comes from the places you visit least frequently.

her PhD thesis research (don't worry, we couldn't see the credit records themselves).[16]

Figure 16 shows the shopping pattern of a typical adult over a month. The bigger circles represent the places one person visited most and the smaller circles show the places they visited least. The arrows illustrate the patterns in which this person moved from place to place, averaged over an entire month. In the *Nature* journal *Scientific Reports*,[17] Coco and I reported together with collaborators Alejandro Llorente, Manuel Cebrian, and Esteban Moro that there is a very regular, lawlike relationship between the number of times various locations are visited, with the most frequent place dominating all the rest combined, the second-most frequent dominating everything but the first, and so on.[18] And, of course, the places visited most frequently were close to home while those visited infrequently tended to be farther away.

The important consequence is that the patterns of shared experiences follow the same general rule as the patterns of social ties. The stores, restaurants, and entertainment places that people visit most are also likely to be visited by their friends and so are unlikely to introduce new ideas into their social network. The greatest chance of an experience that is new to everyone in their social network occurs at the places they visit least frequently. Exploration for new ideas tends to be most fruitful far away, while the common experiences of everyday life are elevated to the status of social norms by engagement within the local community.

Another interesting outcome that emerged when Coco analyzed people's purchasing behaviors was that their patterns of exploration have statistics that are similar to the foraging behavior of animals. We constantly compare among familiar local alternatives to get the best value for our money, of course, but we also go on explorations to find new sources and experiences. These bursts of shopping have the same character as when animals occasionally choose to hunt in a new area, or search for new food sources.

These bursts of exploration—shopping trips, days off that are spent wandering around the city, weekend getaways—seem to be important in growing the local ecology of cities. If we looked at cities with greater than average rates of exploration in the credit card data, we found that in subsequent years they had a higher GDP, a larger population, and a greater variety of stores and restaurants. It makes sense that more exploration, which results in a greater number of interactions between current norms and new ideas, would be a driver of innovative behavior.

Moreover, as cities grow, the ecology of opportunities they offer becomes more complex, just as with biological ecologies. Interestingly, it is in cities that are growing more wealthy that the destinations of exploration grow more common and become overrepresented relative to their frequency in average cities. It seems that not only does exploration result in growing more creative and richer cities, but the process is self-reinforcing. Greater exploration begets greater opportunities for exploration.

Curiosity and exploration: Standard economic theory would predict that people's explorations would diminish as they get to know a neighborhood, figure out the best places to buy things, and discover the purchasing patterns that best suit their lifestyles. But this isn't what happens. Instead, people's exploration is open-ended, and they seem to never stop sampling new stores and services.

Our data show that people are more than simple economic creatures. They do explore in order to find better deals, but they also explore from a sense of curiosity. This tendency is most evident in the wealthiest segments of society. With these people, the rate at which they explore new stores and restaurants is unconnected to the rate at which they switch where and what they buy. They change their patterns of buying at a rate that is similar to most of the population, but their rate of exploration is hugely greater. This suggests that when people have abundant resources, it is their curiosity and social motivations that drive their exploratory behavior and not the desire to find cheaper prices or a better product.

Indeed, when my research group studied the relationships between wealth and social exploration patterns within the Friends and Family community of young families, we found the same patterns.[19] Using data from cell phones and credit card records (see the Reality Mining appendix), we found that both the better-off and the relatively poor groups had roughly the same total amount of face-to-face and phone call socialization. Strikingly, though, the amount of exploratory behavior was consistently greater for the wealthy than for the poor. The difference between these two groups seems to be this: When a family had more money, they changed their balance between contact with familiar people (engagement) and unfamiliar (exploration) in order to obtain greater diversity in the people they interacted with. That is, they used their extra money to increase their exploration.

Importantly, families who used to be wealthy but who now had less money now had lower levels of exploration. So the effect isn't just that the wealthy have traditions of exploration that are different from those of the poor. Instead, families' habits change with the amount of disposable income they have. In fact, the relationship between the amount of disposable income and amount of exploration is very predictable: For each additional dollar of disposable income we see a small increase in both the diversity of socialization and the diversity of store visits. In Chapter 11 we will see that this effect can be used to accurately map the wealth of neighborhoods, since their patterns of explorations are reliable signals of their disposable income.

Unlike what might be expected if the "strength of weak ties" model were true (i.e., the idea that having more social ties results in more wealth), exploration does not seem to translate into greater wealth in the short term. Instead, it is the other way around: Wealth allows people to invest more in exploration. Perhaps this is because good financial status makes people feel more confident and secure in exploring new social opportunities. Exploration appears to be driven by the human need for social contact and novelty rather than by the search for wealth.[20]

The fact that cities with more exploration tend to have greater growth in their wealth suggests that harvesting new experiences and meeting new people does pay off, but it just takes a while. Exploration benefits the city as a whole, and that increase in idea flow within the city is bound to help both individuals and their families, even if only indirectly.

Idea Flow in Cities

Now that we understand more about exploration and networks of social ties within cities, we can ask: Can a city's productivity really be predicted by how far ideas travel and how fast its citizens gain access to new ideas? To test this we need to calculate the rate of idea flow within various cities and compare that to GDP, number of patents, and similar measures of output. The details of this calculation are discussed in the Math appendix.

When we do this calculation we find that the flow of ideas along social networks provides a remarkably accurate account for statistics such as GDP per square mile, as illustrated in Figure 17. The same model gives similarly accurate accounts of patenting rate, R&D investment rate, crime rate, and other features of urban

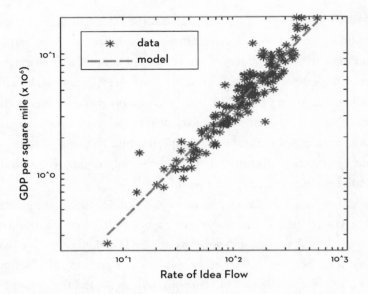

Figure 17. The model of idea flow along social ties accurately predicts GDP per square mile.

life. Idea flow by itself accounts for many of the major features of city life without having to consider additional social structures, such as specialization or classes.

The rate of idea flow is intrinsically a function of the ease of access and interaction between residents living in the same city. There are several factors, however, that can affect the flow of ideas. Consider the example of Beijing, China, which has a very high population density. Due to its traffic jams, however, Beijing is de facto divided into many smaller cities with limited transportation capacity between them. As a result, Beijing does not have as high an idea flow as cities that have a lower population density but better public transportation.

Because of the dependence of idea flow on transportation efficiency, the idea flow equations can be turned around and GDP can be used to calculate the average commuting distance. It turns out that in the United States the average is about thirty miles, whereas in the large cities of the European Union it is about eighteen miles.[21] Both of these numbers are pretty close to the official government statistics, which is surprising given that they were calculated by using only the average structure of social ties across all cities, the cities' population densities, and the measured GDP. In developing countries the average commuting distances are much less, suggesting that these countries could harvest huge gains in productivity and creative output from upgrading their transportation infrastructures.[22]

Designing Better Cities

Traditional theories of city growth emphasize markets and classes, suggesting specialization in industry or new categories of highly trained workers as generative models of city development. In contrast, the social physics approach provides a plausible and empirically grounded model that does not require the presence of these special social structures. Instead, it relies only on the fine-grain characteristics of human social interaction: the distribution of social ties, the flow of ideas along these ties, and the means by which those ideas are converted into new behaviors and new social norms by engagement within peer groups.

In previous chapters I showed how shaping idea flow can improve the productivity of companies. This also provides insight into how we might design better cities. Imagine, for instance, that we wanted to both maintain social norms within civil society and promote innovation within business and the arts. Social physics tells us that if we just increase the population density of the city, or if we just upgrade the transportation network, then we will be promoting both increased crime and increased creative output.[23] But what if we could have both the high levels of social engagement characteristic of traditional villages (and hence their lower crime rate), and the high levels of exploration characteristic of sophisticated business and cultural areas (and hence their greater creative output)?

We want to increase engagement in the residential areas, which will lead to stronger norms of behavior, but *not* increase the amount of exploration for everyone, since that would lead to more crime along with greater innovation. The failure of most city zoning is that if cities segregate by function, then exactly the wrong change in the structures of social ties occurs: Engagement decreases locally (if an area is all just apartment blocks, people rarely get out and meet each other), and exploration increases (since people have to go elsewhere to do anything), and as a consequence the social fabric of the neighborhood is pulled apart. What we want is the opposite: self-contained towns in which people meet each other regularly and there are many friends of friends. As famous urban advocate Jane Jacobs argued, a healthy city has complete, connected neighborhoods.[24]

The best size for such a city can even be calculated: If within each peer group everyone is a friend of a friend, then the math of social physics indicates that we get maximum engagement for populations of up to roughly one hundred thousand people.[25] This suggests that the best solution is small-to-medium-sized towns in which everyone is within walking distance of a town center, the stores, the schools, and the clinics.[26]

For maximum creative output, though, the business and cultural areas should maximize opportunities for exploration. This goal suggests packing as many people as possible into a central city with very efficient and cheap transportation. Ideally, each of our small towns would have a *Star Trek* transporter in the town center that instantly whisks people off to a hot economic center where big multinational corporations have their headquarters and to cultural centers with live events and major museums. The goal is to get maximum exploration in the economic and cultural centers, along with maximum engagement in the towns.

As it turns out, this design is similar to that adopted by Zurich when it was faced with an exploding population. The key was an amazingly extensive, quick, and cheap-to-use light-rail transportation system that enabled people to quickly and conveniently travel into Zurich's city center, thus encouraging them to live in the relatively small and inexpensive towns and villages surrounding it. Many, if not most, people can walk from home to rail stop, travel for fifteen minutes, and then walk the rest of the way to work or to a cultural event.

Today over 60 percent of Zurich's population uses this public transportation system. As a result, exploration and idea flows within the central city are maximized during work hours and cultural events, and engagement is maximized within the surrounding villages. Critically, most people still work in the villages, and it is easy for everyone to attend central cultural events, so that city and village life don't become too separate. The result is that the center of Zurich has the dense flows of new ideas necessary for a flourishing work and cultural environment, but the surrounding villages also have the strong social engagement required to keep them healthy. As a consequence, Zurich has maintained its standing as a world economic center and has been growing into a world-class cultural center, all while maintaining Swiss security and tradition.

Historically this pattern has been repeated in many of the world's best cities. Paris, London, New York, and Boston all were built from small, walking-sized neighborhoods that were later linked by subways and light rail. In some cases this neighborhood structure has been subverted or overwhelmed, but it still remains a source of strength for them.

Urban planners are also beginning to use this type of approach for the renovation of decaying cities. Just as in Zurich, the social physics model suggests that the right approach is to focus on creating a hot inner core, which will have high productivity and creative output. This is the approach planners in Detroit are trying, by working to create a tiny hot new city inside the decaying sprawl of the original one.

Data-Driven Cities

As we have seen in every chapter of this book, social network structure has a dramatic effect on the access to information and ideas.[27] Social tie density is a key determinant behind the flow of ideas between individuals, which in turn determines the spread of new behaviors. Higher social tie density produces greater levels of idea flow, leading to increases in productivity and innovation.

The mathematics of how ideas spread and convert into new behaviors quite precisely accounts for the empirically observed growth of cities across multiple features and different geographies. There is no need to appeal to assumptions about social hierarchies, specialization, or other special social constructs in order to explain how GDP, research and development, and crime grow with increasing city population.

Social physics suggests that the reasons for creating cities are not that different from the reasons for creating work environments such as research parks or universities: We want to engineer the environment to enhance both exploration and engagement. While current digital technology makes remote interaction and collaboration extremely easy and convenient, we have seen that today's digital technology is not as good at spreading new ideas as are face-to-face interactions.

Consequently, the importance of packing people physically close to each other is still critical to promoting greater idea flow.[28] Easy face-to-face access between individuals enhances exploration, engagement, and the rate at which new ideas are converted into behaviors. Thus, physical proximity remains perhaps the major factor in productivity and creative output.

But there is hope for digital communications helping tie together distant groups. High-resolution digital communication can already leverage face-to-face interaction in important ways and may someday become a rich enough communication channel to be just as effective as face-to-face interactions (see Digital Networks versus Face-to-Face special topic box [page 172]). Unfortunately, the convenience of current-day digital communications already makes it easy to create echo chambers in which rumors circulate and the same ideas circulate again and again. If we can begin to keep track of the provenance of ideas, however, then we can also begin to break up echo chambers. As I will argue in the next chapter, this will also be critical to protecting our privacy.

Next Steps

In the last two chapters we have seen how big data and social physics can combine to create a data-driven city. By using the concepts of social physics, it appears that we can make such cities both more productive and more creative, while at the same time minimizing negatives such as crime, excessive energy use, and disease. The recommendations about city structures that come from social physics are similar to those of famous urban advocate Jane Jacobs, but what social physics has added is a quantitative, mathematical basis for the recommendations. By understanding cities as idea engines, we can use the equations of social physics to begin to tune them to perform better.

How can we accomplish these same goals for all of society? In the final section of the book I will discuss how we can go about using our new digital nervous system for the benefit of the entire society, how we can address the tension between privacy and public goods, and, finally, the principles by which we could design a safe, equitable, and secure society.

DIGITAL NETWORKS VERSUS
FACE-TO-FACE

●

People always ask about the role of digital media, such as social networks or phone calls, as compared to face-to-face interactions. The reason behind this interest in digital media is that their low cost and scalability gives rise to the hope of a cheap way to manage companies, influence customers, and reach citizens. The answer is, of course, complicated. The key points to consider are trust and social learning.

Digital media don't convey social signals as well as face-to-face interactions, making it harder for people to read each other, and so digital media are less useful in generating the trust needed for behavior change. In so-called trust experiments, which set one's potential profit by cooperating and trusting others in a group against the possibility of individually profiting from defecting from the group, we find that people who are interacting over digital media almost always defect.[29]

Similarly, when we look at communication channels versus moods, we find that on days when people are either in a really bad mood or a really good mood, they shun e-mail, messaging, and social media, and instead turn more to face-to-face interactions and telephone calls.[30] When we need comfort or are especially happy, we want rich channels of interactions.

In addition, most digital social media are sporadic, asynchronous, and sparse. As we saw in the digital engagement section of Chapter 4, this means that it is difficult to get repeated, frequent exposure to the behavior of trusted peers. As a result, most digital social media are better at spreading facts (and rumors) than spreading new habits. Where it gets complicated is that once a social norm is in place—probably learned through face-to-face interactions—then electronic reminders can be quite effective. For instance, it is real-world interactions that drive most electronic interactions but once begun, electronic media can reinforce a trusted relationship, even though the people remain physically separated.

PART FOUR

Data-Driven Society

•

Data-Driven Societies

WHAT WILL A DATA-DRIVEN FUTURE LOOK LIKE?

We have seen that the digital bread crumbs we leave behind provide clues about who we are and what we want. That makes these personal data immensely valuable, both for public goods and for private companies. As European consumer commissioner Meglena Kuneva said recently, "Personal data is the new oil of the Internet and the new currency of the digital world."[1] This new ability to see the details of every interaction, however, can be used for good or for ill. Therefore, maintaining the protection of personal privacy and freedom is critical to our future success as a society.

A successful data-driven society must be able to guarantee that our data will not be abused—and perhaps especially that government will not abuse the power conferred by access to such fine-grain data. To achieve the positive possibilities of a data-driven society we require what I have called the New Deal on Data—

workable guarantees that the data needed for public goods are readily available while at the same time protecting the citizenry.[2] We must develop much more powerful and sophisticated tools for privacy and reach a consensus that allows us to use personal data to both build a better society and to protect the rights of the average citizen.

A key insight that motivates the creation of a New Deal on Data is that our data are worth more when shared, because they can inform improvements in systems such as public health, transportation, and government. For instance, we have seen that data about the way we behave and where we go can be used to minimize the spread of infectious disease (see Chapter 8). In that example, I described how we were able to use these digital bread crumbs to track the spread of influenza from person to person on an individual level. And if we can see it, we can stop it. In this instance, the result of sharing personal data is that we can build a world where the threat of infectious pandemics is greatly diminished.

Similarly, for those worried about global warming, these shared, aggregated data now show us how patterns of mobility relate to productivity (see Chapters 8 and 9). In turn, this provides us with the ability to design cities that are more productive and at the same time more energy efficient. But in order to be able to obtain these results and make a greener world, we need to be able to see the people moving around; this depends on many people being willing to contribute their data, even if only anonymously and in aggregate.

Unfortunately, today most personal data are siloed off in private companies and therefore largely unavailable. Private organizations collect the vast majority of personal data in the form of location patterns, financial transactions, phone and Internet communications, and so on. These data must not remain the exclusive domain of private companies, because then they are less likely to contribute to the common good. Thus, these private organizations must be key players in the New Deal on Data's framework for privacy and data control. Likewise, these data should not become the exclusive domain of the government, because this will not serve the public interest of transparency, and we should be suspicious of trusting the government with such power.

While concrete examples such as better health systems and more energy-efficient transportation systems motivate a New Deal on Data, there is an even greater public good that can be achieved by efficient and safe data sharing. As we have seen in Chapters 5 and 9, greater idea flow results in greater productivity and creative output. In the long run it is the creative output of our society that will lift living standards and enable more meaningful lives. Therefore, a key goal of a New Deal on Data should be the promotion of greater idea flow.

One way to enhance idea flow is through the creation of a public data commons, e.g., freely available maps and statistics about matters such as employment and crime rates. Robust data sharing and anonymization technology can create a data commons that respects citizens' privacy, corporations' competitive interests, and, in addition, provides oversight of government.[3] At the end of Chapter 11, I will discuss what may be the world's first large-scale digital commons, and explain how a resource such as this can be used to help build a better society.

But not all personal data can reside in a public commons; we will want a great deal of personal data to remain private. To enable sharing of personal data and experiences, we also need secure technology and regulations that enable individuals to safely and conveniently share personal information with each other, with corporations, and with government. Consequently, the heart of a New Deal on Data must be to provide both regulatory standards and financial incentives that entice owners to share data while at the same time serving the interests of both individuals and society at large. We must promote greater idea flow among individuals not just corporations or government departments.

A New Deal on Data

It has long been recognized that the first step to promoting liquidity in land and commodity markets is to guarantee ownership rights so that people can safely buy and sell.[4] Similarly, the first step toward creating greater idea flow ("idea liquidity") is to define ownership rights.[5] The only politically viable course is to give individual citizens the rights over the data that are about them, and in fact, in the European Union these rights flow directly from the constitution. We need to recognize personal data as a valuable asset of the individual that is given to companies and government in return for services.[6]

The simplest approach to defining what it means for an individual to own their own data is to draw an analogy with the English common law ownership rights of possession, use, and disposal:

- *You have the right to possess data about you.* Regardless of what entity collects the data, the data belong to you, and you can access your data at any time. Data collectors thus play a role akin to a bank managing the data on behalf of its customers.
- *You have the right to full control over the use of your data.* The

terms of use must be opt-in and clearly explained in plain language. If you are not happy with the way a company uses your data, you can remove it, just as you would close your account with a bank that is not providing satisfactory service.

- *You have the right to dispose of or distribute your data.* You have the option to have data about you destroyed or deployed elsewhere.

Individual rights to personal data must be balanced with the needs of corporations and governments to use certain data—account activity, billing information, and so on—to run their day-to-day operations. This New Deal on Data therefore gives individuals the right to possess, control, and dispose of copies of these required operational data, along with copies of the incidental data collected, such as location and similar context. Note that these ownership rights are not exactly the same as literal ownership under modern law, but the practical effect is that disputes are resolved in a different, simpler manner than would be the case for (as an example) land ownership disputes.

In 2007, I first proposed the New Deal on Data to the World Economic Forum. Since then, this idea has been run through various discussions and eventually helped shape the 2012 Consumer Data Bill of Rights in the United States, along with a matching declaration on Personal Data Protection in the EU. These new regulations are intended to accomplish the combined trick of breaking data out of the silos they are currently held in, thus enabling public goods, while at the same time giving individuals greater control over data about themselves. But, of course, this is still a work in progress, and the battle for individual control of personal data continues.

Enforcement

How can we enforce this New Deal? The threat of legal action is insufficient, because if abuses can't be seen, then they can't be prosecuted. Plus, who wants more lawsuits anyway?

The current best practice is a system of data sharing called trust networks. Trust networks are a combination of a computer network that keeps track of user permissions for each piece of personal data, and a legal contract that specifies both what can and can't be done with the data, and what happens if there is a violation of the permissions. In such a system all personal data have attached labels specifying what the data can and cannot be used for. These labels are exactly matched by terms in a legal contract between all the participants stating penalties for not obeying the permission labels and giving the right to audit the use of the data. Having permissions, including the provenance of the data, allows automatic auditing of its use, and enables individuals to change their permissions and even withdraw the data.

A system like this has made the interbank money transfer system among the safest systems in the world, but until recently such technology was only for the big guys. To give individuals a similarly safe method of managing personal data, my research group and I here at MIT, in partnership with the Institute for Data Driven Design (cofounded by John Clippinger and myself),[7] have helped build openPDS (open Personal Data Store), a consumer version of this type of system, and we are now testing it with a variety of industry and government partners.[8] Soon sharing personal data could become as safe and secure as transferring money between banks. This system is described in more detail in the openPDS appendix.

The Wild, Wild Web

So far I have emphasized the new sensor-derived sources of personal data because the extent and nature of these data are unfamiliar to many people. Of course, massive amounts of personal data are already on the Web. Most of it consists of information contributed by users to social network sites, blogs, and forums; transaction and registration data from online merchants and organizations; and browsing and click-stream histories. Companies are just starting to reality mine user-contributed images and video that, although consciously contributed, present many of the same dangers of inadvertent harm as passively collected sensor data such as call records and phone location data.

The Web has evolved in an unregulated environment with no coherent privacy standards about personal data. Consequently, the rights to such data are unclear and vary from site to site. In contrast, mobile phone, medical, and financial data are collected by heavily regulated industries with fairly clear ownership rules, and these data could be made more widely accessible by marrying a New Deal on Data to their existing frameworks, thus enabling carefully controlled sharing of these sorts of personal information.

But what of the Wild, Wild Web? Fortunately, existing Web companies are coming under pressure to conform to the higher standards being imposed on regulated industries. Perhaps the best example is Google, a participant in the World Economic Forum Rethinking Personal Data initiative that I help lead. Following the initial round of discussions at the World Economic Forum, the company released Google Dashboard (www.google.com/dashboard), which lets users know what data it has about them. After a second round of discussions, the company formed the Data Liberation Front (www.dataliberation.org), a group of Google engineers whose mission statement says that "users should be able to control the data they store in any of Google's products" and whose goal is to "make it easier to move data in and out." When my former student Bradley Horowitz helped to launch Google+ in June 2011, data ownership and portability was a key design element. These steps toward individual control of personal data are just a start, but the pressure is building for all companies to fully embrace a New Deal on Data.

Data-Driven Systems: Challenges

The ability to safely share data will inevitably produce governance and policies that are more driven by data. We can hope to achieve much better social outcomes through the use of big data and social physics analysis. Perhaps just as important, social physics enables us to use big data and visualizations to get near to real-time insight into how our policies are performing, and this greater transparency can help the public have meaningful control of how and when policies should be adjusted and revised.

As an example, in my laboratory we are now building a Web tool based on Google maps, but instead of just showing roads and satellite imagery, it shows maps of poverty, infant mortality, crime rate, change in GDP, and other social indicators, updated daily and on a neighborhood-by-neighborhood basis. Using this new mapping ability, it can quickly be seen where new government initiatives are working and where they are not.[9]

The biggest barrier to building better societal systems using such massive data, however, is not their size or speed, nor even privacy and accountability in sharing. Instead, the biggest challenge is learning how to build social institutions based on the analysis of billions of individual connections. We need social physics, so that we can move from systems based on averages and stereotypes to ones based on the analysis of individual interactions.

M oving beyond the closed laboratory: Our traditional methods of testing and improving government, organizations, and so on are of limited use in building a data-driven society. Even the scientific method as we normally use it no longer works, because there are so many potential connections that our standard statistical tools generate nonsense results.

The reason is that with such rich data, it is easy to be misled by spurious correlations. For instance, let's imagine we discover that people who are unusually active are more likely to get the flu. This is a real example: When we examined the minute-by-minute behavior of a small university community—a real-time flow of gigabytes per day for an entire year—we noticed that an unusual level of running around often predicted the onset of the flu. But if we can only analyze the data using traditional statistical methods, we have a problem in understanding *why* it is true. Is it because flu virus makes us more active in order to spread itself more quickly? Or did interacting with many more people than usual make them more likely to catch it? Or is it something else? From the real-time stream of data by itself, we just can't know.

The point here is that normal analysis methods don't suffice to answer these sorts of questions because we don't know all the possible alternatives and so we can't form a limited, testable number of clear hypotheses. Instead, we need to devise new ways to test the causality of connections in the real world. We can no longer rely on laboratory experiments; we need to actually do the experiments in the real world, and usually on massive, real-time streams of data.

Using live data to design institutions and policies is outside of our normal way of managing things. We live in an era that builds on centuries of science and engineering, and the standard choices for improving systems, governments, organizations, and so on are fairly well understood. Therefore, our scientific experiments normally need only consider a few clear alternatives (i.e., plausible hypotheses).

But with the coming of big data we are going to be operating very much out of our old, familiar ballpark. These data are often indirect and noisy, and so interpretation requires greater care than usual. Even more important, a great deal of the data is about human behavior, and the questions are ones that seek to connect physical conditions to social outcomes. Until we have a solid, well-proven, and quantitative theory of social physics, we won't be able to formulate and test hypotheses in the simple, clear-cut manner that today allows us to reliably design bridges or test new drugs.

Therefore, we must move beyond the closed, laboratory-based question-and-answer process that we currently use and begin to manage our society in a new way. We have to begin to test connections in the real world far earlier and more frequently than we have ever had to do before, using the methods my research group and I have developed for the Friends and Family or the Social Evolution studies. We need to construct living laboratories—communities willing to try a new way of doing things or, to put it bluntly, to be guinea pigs—in order to test and prove our ideas. This is new territory, and so it is important for us to constantly try out new ideas in the real world in order to see what works and what doesn't.

An example of such a living lab is the "open data city" I

have just help launch within the city of Trento in Italy, along with Telecom Italia, Telefónica, the research university Fondazione Bruno Kessler, the Institute for Data Driven Design, and local companies. Importantly, this living lab has the approval and informed consent of all of its participants—they know that they are part of a gigantic experiment whose goal is to invent a better way of living. More detail on this living lab can be found at http://www.mobileterritoriallab.eu.

Its goal is to develop new ways of sharing data to promote greater civic engagement and exploration. One specific aim is to build upon and test trust-network software such as our openPDS system.[10] Tools such as openPDS make it safe for individuals to share personal data (e.g., health data, facts about your children) by controlling where your data go and what is done with them.

The specific research questions we are exploring depend upon a set of personal data services designed to enable users to collect, store, manage, disclose, share, and use data about themselves. These data can be used for the self-empowerment of each member or (when aggregated) for the improvement of the community through a commons that enables social network incentives. The ability to share safely should enable better idea flow among individuals, companies, and governments, and we want to see if these tools can in fact increase productivity and creative output on the scale of an entire city.

An example of an application enabled by the openPDS trust framework is the sharing of best practices among families with young children. How do other families spend their money? How much do they get out and socialize? Which preschools or doctors do people stay with for the longest time? Once the individual gives permission, our openPDS system allows such personal data to be collected, anonymized, and shared with other young families safely and automatically.

The openPDS system lets the community of young families learn from each other without the work of entering data by hand or the risk of sharing through current social media. While the Trento experiment is still in its early days, the initial reaction from participating families is that these sorts of data-sharing capabilities are valuable, and they feel safe sharing their data using the openPDS system.

The Trento Living Lab is letting us investigate how to deal with the sensitivities of collecting and using deeply personal data in real-world situations. In particular, the lab will be used as a pilot for a New Deal on Data and for new ways to give users control of the use of their personal data. For example, we will explore different techniques and methodologies to protect the users' privacy while at the same time being allowed to use their personal data to generate a useful data commons. We will also explore different user interfaces for privacy settings, configuring the data collected, and the data disclosed to applications and shared with other users, all in the context of a trust framework.

Challenge to human understanding: A second challenge in building a data-driven society is human understanding. With the arrival of dense, continuous data and modern computation, we can now map out the details of society and build mathematical models of them. But these raw mathematical models are very far beyond most humans' understanding. They have too many variables and the relationships are too complex for the poor human mind. While these highly detailed and very mathematical models are good for building automatic systems for traffic, electric power, and the like, they are almost useless for guiding individual human decision making.

In order for governments and citizens to make decisions about our society, we need to build a human-scale, intuitive understanding of social physics. I believe that there needs to be a dialogue between our human intuition and the big data statistics, something that is not built into most of our management systems today. Today most people have little concept of how to use big data analytics, what they mean, and what to believe. A new language, one that goes beyond markets and classes and captures how detailed connections between people determine change, will help us develop this understanding. My hope is that the language and concepts in this book will let us bridge this gap.

Social Physics versus Free Will and Human Dignity

Some people react negatively to the phrase social physics, because they feel that it implies that people are machines without free will and without the ability to move independently of our role in society. The social physics I envision, though, acknowledges our human capacity for independent thought but does not need to try to account for it. Social physics is based on statistical regularities that span the population, i.e., things that are true of almost everyone almost all of the time.

Our individual, conscious system of beliefs is formed by deductions from facts and assumptions, and this allows us to deduce an entire world of conclusions. If we change even just one core fact, assumption, or rule, however, then our entire system of beliefs can switch dramatically. Nor is this fragile generativity just a theoretical possibility; such profound changes often occur when people go through army boot camp or are inducted into religious cults. In such cases, a person's entire system of beliefs can change in the course of just a few days or weeks. A social physics based on regularities shared by all people cannot account for the sort of individual fluidity displayed by our belief systems.

Instead, the power of social physics comes from the fact that almost all of our day-to-day actions are habitual, based mostly on what we have learned from observing the behavior of others. Because most of our actions are habitual and based on physical, observable experiences, i.e., stories heard, actions seen, etc., they can be described as repeated patterns. This means that we can observe humans in just the same way we observe apes or bees and derive rules of behavior, reaction, and learning.

Unlike apes or bees, however, we know that humans always have an internal, unobservable thought process, and this will occasionally emerge to defeat our best social physics models. The consequence is that although we can use social physics to design living spaces, transportation systems, and governments that are tuned for daily routine and typical human behavior, we will always have to leave room for unusual personal choices. What is surprising is that the data tell us that deviations from our regular social patterns occur only a few percent of the time. As a consequence, we have to be very careful to provide for these green shoots of individual innovation and not give in to arguments about cost and only support the most common patterns. (Please see the Fast, Slow, and Free Will appendix for more detail on this subject.)

Because modern culture puts so much emphasis on independence and personal choice, it is often difficult to realize that it is a *good* thing that most of our life is highly patterned, and that we are all quite similar rather than being completely different individuals with different patterns of behavior. The fact that most of our attitudes and thoughts are based on integrating the experiences of others is the very basis for both culture and society. It is why we can cooperate and work together toward common goals.

There is another reason why people should prefer the concepts of social physics to that of markets and classes. Because markets and classes are averages or stereotypes, reasoning that uses these terms leads inevitably to considering all the people in the market or class to be the same. Adam Smith's markets end up being as dehumanizing as Karl Marx's classes.

All of this has practical consequences and is more significant than word choice preferences. Because everyone understands the logic of markets and class competition and there is no scientific, workable alternative readily available, we too often describe society as a continual competition and categorize people primarily by their class or market membership. We think of people as being a trendy millennial, an upscale baby boomer, or a white Republican. This way of thinking leads naturally to stereotyping and an over-emphasis on easily measured features such as money or fame. It leads to a winner-take-all popular culture, militant capitalism, and governments that rely too much on competition and market incentives in order to manage society.

Having a mathematical, predictive science of society that includes both individual differences and the relationships between individuals has the potential to dramatically change the way government officials, industry managers, and citizens think and act. For instance, it can allow them to use the tools of social network incentives in order to establish new norms of behavior, rather than relying on regulatory penalties and market competition. Imagine government decision making by combining the technology of the "Red Balloons" contest (Chapter 7) and that of the wisdom of the crowd (Chapter 2) to recruit tens of millions of people to discover solutions and enlist support within millions of local town hall meetings—it is possible, and could be systematically better than today's decision mechanisms. To accomplish this change we need a language and logic that everyone can understand and that has proven its ability to be more useful than the old language of markets and classes. I believe that the language of social physics— exploration, engagement, social learning, and measurement of idea flows—has the potential to serve that role.

Design for Harmony

HOW SOCIAL PHYSICS CAN HELP US DESIGN A HUMAN-CENTRIC SOCIETY

Today the majority of societies around the world are based on free markets, although almost always with some fixes and restrictions. This model for society has its roots in eighteenth-century notions of natural law: the idea that humans are self-interested and self-commanded and that they relentlessly seek to gain from the exchange of goods, assistance, and favors in all social transactions. Open competition for such theoretical individuals is a natural way of life, and if all the costs (e.g., pollution, waste) are taken into account, then the dynamics of open competition can result in an efficient society. As Adam Smith explained:

> They are led by an invisible hand to make nearly the same distribution of the necessaries of life, which would have been made,

had the earth been divided into equal portions among all its
inhabitants, and thus without intending it, without knowing it,
advance the interest of the society, and afford means to the mul-
tiplication of the species.[1]

The power of markets to distribute resources efficiently, together with the assumption that humans are relentless competitors,
is the bedrock of most modern societies. It works well for stocks and
commodities, and reasonably well for wages and housing, and increasingly the trend is to apply market thinking to all sectors of society. But does this eighteenth-century understanding of human
nature truly form a good model for all of these sectors of our modern society? I think not.

Competition versus cooperation. As the previous chapters of
this book have demonstrated, one major flaw in this view
of human nature is that people are not simply self-interested, self-
commanded individuals. What we are interested in, and our com-
mand mechanism itself, is overwhelmingly determined by social
norms created by interactions with other people.

Modern science now understands that cooperation is just as
important and just as prevalent in human society as competition.[2]
Coordination and cooperation among peers are shaping forces that
are very powerful—our friends watch our backs, in sports and business teammates cooperate to win against other teams, and everywhere people support family, children, and the elderly. In fact, the
whole concept of shared culture and cultural norms is based on
the coordination of individual behaviors. Let us look a little more
closely at the role of cooperation in modern society and how the

fact of cooperation contrasts with the idea that people are relentless competitors.

As we have seen in previous chapters of this book, people cooperate with each other to establish social norms. These norms are what we call culture. In fact, the main source of competition in society may not be among individuals but rather among cooperating groups of peers. Further, for each opportunity or threat, the relevant group of peers is a different set of individuals. For example, London bankers coordinate with each other to make money, using shared strategies and shared standards within the industry. Similarly, New York lawyers have shared norms that let them thrive as a group within their local ecosystem, and politicians create shared traditions and methods about how to trade money interests against citizen interests, while at the same time courting journalists. In each of these cases it is cooperation—implicit and explicit agreements about how to coordinate our behavior with our peers—that defines competitive interactions with the rest of society.

Classes versus peer groups. Peer groups with shared norms are different from the traditional idea of class, because they are not defined only by standard features such as income, age, or gender (e.g., traditional demographics), their skills and education (per Max Weber),[3] or their relationship to the means of production (per Karl Marx).[4] Instead, group members are peers in the context of a particular situation. In one situation a person's peer group may be people with the same hobby (e.g., choral singing), in another, a similar history (e.g., same graduating high school class), and in yet another context it may be people with the same job description (e.g., firefighters). Therefore, while a person is a member of only one traditional class, that person is also a member of many different peer groups. Within each peer group, members learn from each other and so create a shared common sense that is different depending on which hobby they enjoy, where they graduated from, or where they work.

Nor are these peer groups simply cases of economic coalitions working together, because these groups also foster strong norms about a wide range of topics, including life goals, moral values, and even clothing. Members develop an entire culture, a lifestyle that then leaks into the other peer groups that they are part of. When the bankers get home, they are mothers or church leaders, and some of the banker culture rubs off on those groups, and vice versa. Typically, no one is defined only by their job; people who come close to being that one-dimensional are considered odd, and perhaps slightly unhinged.

From this point of view, political or economic labels such as bourgeoisie, working class, Democrat, or Republican are often inaccurate stereotypes of groups of people who actually have widely varying individual characteristics and desires. As a result, reasoning about society in terms of classes or parties is imprecise and can lead to mistaken overgeneralizations. In the real world, a group of people only develops deeply similar norms when they have both strong interactions and they recognize each other as peers.

Markets versus exchanges. Just as classes are oversimplified stereotypes of a fluid and overlapping matrix of peer groups, the idea of a market is a similarly flawed idealization, in this case in which it is imagined that all the participants can see and compete evenly with everyone else. In reality, some people have better connections, some people know more than others, and some purchases are more difficult than others, due to distance, timing, or other secondary considerations. A basic example is found in today's stock market. There the average person has much less information than a professional stock trader, and even sophisticated, professional traders may be at a disadvantage with computerized high-frequency traders who react to price changes in mere milliseconds. Our ideal of a free market suddenly turns out to be more complicated.

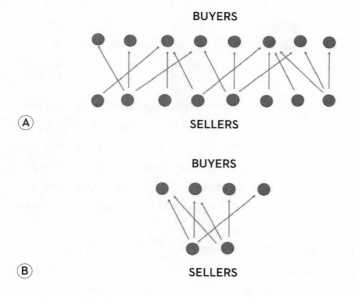

Figure 18: (a) a classical market, (b) an exchange network. An exchange network is a market where trade options are limited to connections within the social network. Trust and personalized service is much more likely to develop within an exchange network.

A more important example of how situations that look like traditional markets are often really something else is illustrated in Figure 18. In Figure 18(a), we have a depiction of the classic idea of a market. Here a large number of buyers purchase from a large number of sellers, and so prices settle down to be efficient and uniform across the economy. This type of symmetric market is also robust. If one seller has a problem, that is, their supplies run out or their delivery truck breaks down, then the other sellers make up the shortfall.

The real world, however, looks more like the exchange network illustrated in Figure 18(b), in which buyer-seller relationships are more constrained and asymmetric. By analyzing the U.S. government data about which companies buy from which suppliers, my colleagues Daron Acemoglu, Vasco Carvalho, Asu Ozdalgar, and Alireza Thabaz-Salehi have shown that most intersector relationships in the U.S. economy are like the constrained, asymmetric relationships illustrated in Figure 18(b).[5]

An advantage of this sort of constrained trade network is that buyers are more likely to develop stable, trusted relationships with the sellers. With increased stability and trust comes the increased ability to exert social pressure and so sellers can end up customizing their offerings for each particular buyer. This may be why the U.S. economy looks like 18(b) and not 18(a): People prefer trusted and personalized relationships.[6]

These same constraints and asymmetries can also pose a danger, however, because if one large buyer or seller has a problem, then the problem can cascade, hurting all the buyers and sellers they have relationships with. A famous recent example of this was when the president of the Ford Motor Company went to the U.S. Congress to argue for rescuing his fiercest competitor, General Motors. Why? Because Ford and GM depend on many of the same suppliers: If GM went broke, then the suppliers would go broke, and then Ford would be unable to manufacture its own cars. This sort of cooperation between competitors is definitely not something we would expect if we relied on only classic market thinking.

What these examples and the economic data show is that a basic assumption within classic market thinking—that there are many sellers and buyers that can be substituted for each other easily—does not apply to much of the U.S. economy. Instead, we need to think of the economy as a complex network of specific exchange relationships.

Natural Law: Exchanges, Not Markets

Modern society is based on the idea that markets can distribute resources efficiently and on the assumption that humans are relentless competitors. But as we have seen, this is simply not a good description of how our society lives and functions.

Was this ever a good description of human society? That is, was there a time in our human history when we were all fierce competitors in an open fight for resources? While much of our mythology and romantic fiction suggests that this was true of early human societies, science tells us a different story.

Anthropologists report that in the most remote and untouched societies they have found social traditions that are very egalitarian, with surprisingly equal sharing of food and often with distributed, expertise-dependent authority.[7] Physical mobility is limited in these societies, though, so encounters with outsiders are relatively rare. And even when outside encounters do happen, the ability to transmit ideas and information, or even to broker the trade of food and other goods (i.e., to arbitrage between supply and demand), is quite limited due to the lack of writing, sophisticated language, or numeracy.

Of importance for our purposes, though, is that this implies that our minds and cultures evolved in an era when both goods and ideas propagated through individual interactions, and it took a long time for a new idea or valuable goods to spread through the population. In other words, many early societies operated much more like an exchange network than a market: There was no market mechanism or price-setting authority for establishing the value of goods or ideas. Limited mobility meant that supply and demand were limited to exchanges with at most a very few people at any one time, and reputations were mostly earned one-on-one rather than shared through some central authority.

In Ankur Mani's PhD studies with me we used game theory to mathematically examine the properties of exchange networks that were typical of early human societies.[8] Specifically, we asked if these early societies had the same properties as market-based societies or if they were different. In solving the equations, Ankur found that in a society based on exchange networks, Adam Smith's invisible hand functioned locally within the network, and as a bonus, it required no external reputation mechanisms or referees.[9] Moreover, in some important respects a society of traders is better than a society of competitors. Like markets, exchange networks are able to distribute goods equitably, but they also offer better support for individual members and can be more robust in response to outside shocks.

The central reason exchange networks are better than markets is trust. Relationships in an exchange network quickly become stable (we go back again and again to the person who gives us the best deal), and with stability comes trust, i.e., the expectation of a continued valuable relationship. This is different than in a typical market, where a buyer may deal with a different seller every day as prices fluctuate. In exchange networks, buyers and sellers can more easily build up the trust that makes society resilient in times of great stress. In markets, one must usually rely on having access to an accurate reputation mechanism that rates all the participants, or to an outside referee to enforce the rules.

As a consequence of greater stability and trust, the equations showed that the dynamics of exchange networks intrinsically cause them to evolve to be fair, and the surplus generated by the relationship is equally divided between the individuals involved.[10] And as a consequence of more fairness, more stability, and greater levels of trust, exchange networks are also more cooperative, robust, and resilient to outside shocks.[11] That is a good recipe for building a society that will survive.[12]

Adam Smith thought that the invisible hand was due to a market mechanism that was constrained by peer pressure within the community. In the succeeding centuries we have tended to emphasize the market mechanism and forgotten the importance of the peer pressure part of his idea. Our results strongly suggest that the invisible hand is more due to the trust, cooperation, and robustness properties of the person-to-person network of exchanges than it is due to any magic in the workings of the market. If we want to have a fair, stable society, we need to look to the network of exchanges between people, and not to market competition.

These mathematical analyses paint a very nice picture of early human society, and perhaps this is why some of the early societies studied by anthropologists were so stable and egalitarian. Egalitarian and stable, though, does not necessarily mean peaceful. Some of them were very violent, with intertribal warfare playing a very significant role in determining both life expectancy and mixing of the gene pool. I believe that this violence is, at its heart, due to very low levels of idea flow: high levels of engagement within a community combined with low levels of exploration outside the community usually leads to rigid and insular societies. Insular communities (including the society of Adam Smith) often inflict terrible damage on weaker communities with whom they share resources, as I described in the Subjugation and Conflict section of Chapter 4.[13]

But how does this idea of an exchange society apply to modern life? Today we have mass media to spread information, and our much higher levels of mobility allow us to interact with many more people. Do the facts that information is so universally available and our social networks are so broad mean that we have transitioned from an exchange society to a market society?

I think the answer is no. Even though we now have much greater breadth and rate of interaction, our habits still depend mostly on interactions with a few trusted sources—those people whom we interact with frequently—and for each person the number of such trusted individuals remains quite small. In fact, the evidence is that the number of trusted peers that we have today is pretty much the same as it was tens of thousands of years ago.[14]

This small, relatively stable network of trusted peers still dominates our habits of eating, spending, entertainment, political behavior, and even technology adoption, as I described in Chapter 3. Similarly, face-to-face social ties drive output in companies and account for the productivity and creative output of the largest cities (see my group's research studies in Chapters 5 and 9). This means that the spread of new behaviors throughout society is still dominated by local, person-to-person exchanges even in the presence of modern digital media and modern transportation. We still live in an exchange society, albeit one with much greater levels of exploration.

Design for a Networked Society

How are we to use these insights about human nature—the importance of social learning and social pressure, as well as the idea that human society is more of an exchange network than an open market—to design a society better suited to human nature? Social physics suggests that the first step is to focus on the flow of ideas rather than on the flow of wealth, since the flow of ideas is the source of both cultural norms and innovation. Economic theory still provides a useful template for designing the flow of ideas within society, but we have to begin with a more accurate notion of human nature. Because we are not just economic creatures, our models must include a broader range of human motivations, such as curiosity, trust, and social pressure. We must also take into account both the social and dynamic network natures of human society. This means that our focus should be on providing the idea flow required for individuals to make correct decisions and develop useful behavioral norms.

I believe that there are three design criteria for our emerging hypernetworked societies: social efficiency, operational efficiency, and resilience. Let us look at each of these in turn and then ask how they might apply to governments and society more generally.

Social efficiency: In the language of economics, social efficiency refers to the optimal distribution of resources throughout society—a process that, as Adam Smith famously described, occurs through the workings of an invisible hand. Of course, as we saw in Chapter 4, the invisible hand doesn't work unless everyone is engaged in the same social fabric so that peer pressure can ensure that everybody will follow the same set of rules.

When such an inclusive social system is also socially efficient, this means that when one person benefits, the entire society benefits. The reverse is also implied: what harms an individual is likewise bad for society. When most people are sufficiently well off, then the measure of how well a society divides its wealth can be evaluated by the condition of its poorest and most vulnerable members.[15]

Given the well-known shortcomings of human nature, social efficiency is a desirable goal. Applying this principle to the flow of ideas within a society, we see that the exchange of ideas and information between people must reliably provide value not only to the individual but to the whole system.[16]

The traditional way to accomplish the goal of social efficiency has been the open market approach, that is, to provide open, public data in support of fair markets. This is a solution that has dominated our thinking for the last century. While our reliance on open data has provided transparency in many civil systems, the amount and richness of publicly available data are now leading to concerns about the "end of privacy." We have discovered that simple anonymization of personal data just doesn't work reliably, because by combining different data sets people can often be reidentified.

And so we find ourselves in a situation where the guys with the biggest computers can pretty much track everything we do and where we go, creating the danger that we are moving toward a big brother society. Corporations and governments have computational capabilities far beyond what is available to individuals, and this imbalance is quickly becoming a major source of social inequality. The combination of these two trends—greater data access and greater computing power—produces an incredible concentration of power in the hands of government and large corporations.

An alternative to the open market approach that also achieves social efficiency is the exchange network. This approach to sharing ideas and information relies on the strong control of personal data so that they are only ever shared as part of an agreed-to exchange, and that the data never flow any farther. By setting up trusted digital exchange networks rather than using open-market mechanisms, we can control where personal data go and how they are used. As I explained previously, I believe that exchanges on these sorts of networks can more effectively enable Adam Smith's invisible hand, as well as provide increased levels of fairness, trust, and stability.

To illustrate the idea of a trusted exchange network, consider the typical urban experience. In your daily life you have routine interactions with many people: buying coffee, getting a bus ride, etc. You probably don't even know the names of many of these people, and you almost certainly don't know their family members, friends, and colleagues, or what they do when they are not working. Because you interact with them every day, however, these exchanges are trustworthy. That is, you expect that the coffee you get today will taste like it did yesterday, and the price will be about the same.

The fact that you know these "familiar strangers" but not their networks of exchanges means that collusion against them is difficult. As a result, the exchange is safe from many forms of fraud and abuse. Similarly, digital trust networks such as our openPDS system can make information exchanges open and fair while limiting people's exposure to risk through digital filters that provide strong control of personal data. These digital mechanisms make sure that in any exchange individuals share only the minimum personal data that are required, and that the data are used only for the intended purpose.

Today there are long-standing versions of trust networks that have proven to be both secure and robust. As mentioned in Chapter 10, the best-known example is the SWIFT network for interbank money transfers, and its most distinguishing feature is that it has never been hacked.[17] When asked why he robbed banks, bank robber Willie Sutton famously said, "Because that's where the money is." Well, in today's world, the SWIFT network is where the money is—trillions of dollars per day. This trust network has not only kept the robbers away, but it also makes sure the money reliably goes where it is supposed to go.

We can adapt this trust network technology for everyday, person-to-person interactions and so create an exchange network society instead of having to always resort to open-market mechanisms. Just as the banks sign up to the SWIFT network so that they can safely interact with other banks, individuals could sign up for trust networks so that they could safely interact with other individuals or companies, secure in the knowledge that their personal data would be used only in the ways they agreed to.

A trust network, with its emphasis on one-to-one exchanges and strong personal control of data, gives us social efficiency, along with fairness and stability, as properties that are inherent to exchange networks. As the familiar stranger example above illustrates, exchange societies can even feel more natural than the open competition environment championed during the Enlightenment. Perhaps this is because exchange societies seem to be the sort of environment in which the human mind developed and, if so, then exchange networks would be expected to fit especially well with our social instincts and fast reasoning capabilities.[18]

The open market and strong personal control models are but two approaches to social efficiency. Blends of these two models are also possible. For instance, we could create a limited data commons that is free and open to the public but yields much a greater benefit when combined with personal, private data.

Health care is a good example of such a data commons. Governments are now beginning to force hospitals and drug makers to make information about the effectiveness of their treatments freely available. This public information, combined with the private information in our personal health records, can help us get much better health care. By creating a public data commons that provides depth and context to personal data, we can make our personal data much more useful while at the same time achieving the goal of social efficiency and the equitable flow of information and ideas. I will explore this topic further in the next section of this chapter, by examining what is probably the world's first big data commons: Data for Development (D4D).

Operational efficiency: In addition to social efficiency, we also need operational efficiency. In other words, the infrastructure of our society should work quickly, reliably, and without waste if our society is to thrive in our modern, resource-limited world. In particular, data systems should attempt to provide optimal operation for day-to-day activities, particularly when they are being used to control society's physical networks and systems. Using this definition, our current financial, transportation, health, energy, and political systems all seem to be failing us. Perhaps, in part, this is because they were all designed in the 1800s, when builders relied on rigid, centralized control because the primary sensing and data systems of the time were literally people riding around in horse-drawn buggies.

One step toward achieving this goal of operational efficiency is to create a public data commons that lets us see the big picture in real time. Not every piece of data needs to be part of this god's-eye view of the world, however. The commons generally needs only aggregate anonymous data that are relevant to the tasks at hand. Such aggregate data can be used to set the broad policies and regulate physical social systems, with others using private data to tune the systems through individual private exchanges. An example of this sort of commons-based regulation would be aggregating anonymous medical records (something that must be done with careful legal restrictions and auditing) and analyzing them to discover which drug treatments work best and which drug interactions are dangerous. These aggregates are ideas (e.g., context, action, expected outcome) that can be used to tune the medical treatment of individuals.

While scientists are now learning how to improve our health, transportation, and other public systems through the use and analysis of these sorts of data commons, a missing part of the picture is how to get people to adopt the ideas that may be discovered. Designing an optimal system is useless unless it fits our human natures, because otherwise people won't cooperate, either ignoring or misusing the system.

Social physics has a role both in enabling people to discover the best ideas and then getting them to cooperate. In previous chapters we saw how social networks can provide more effective incentives to promote the development and enforcement of useful social norms. We now need to begin applying these lessons to reinvent our current economic, government, and work systems. Just as with the health and energy conservation experiments described in Chapter 4 and the social network interventions described in Chapter 8, we can begin to use social physics in order to improve the operational efficiency of our social systems.

The key to better systems is real-time monitoring of conditions, continuous exploration for the best response ideas, and then engagement around these in order to obtain a coordinated, consistent response to changed conditions. If we think of the search for good strategies on the eToro platform in Chapter 2, the social pressure for cooperation that we saw in the energy conservation examples in Chapter 4, and the rapid recruitment of participants in the Red Balloon example in Chapter 7, it seems that future systems might look a lot like Wikipedia but founded on overlapping clusters of buddies who have face-to-face relationships rather than being completely virtual and digital. In other words, exploration for good ideas would happen in the digital realm, but engagement for consensus would primarily happen face-to-face. By iterating between exploration and engagement among and between the different groups of buddies, we might be able to scale up the ancient decision-making processes that we see in social species ranging from bees to apes and that are still necessary to win consensus among fast- and slow-thinking humans.

Resilience: A third design principle, resilience, relates to the long-term stability of our social systems. Today's social systems—finance, government, and work—seem to periodically seize up, fall apart, or crash and burn. It is important that we design new systems in which such systemwide failures are less likely to happen. Similarly, social systems that cannot respond quickly and accurately to changing conditions and threats are inadequate for humanity's modern needs. Clearly, our long-term resilience depends on our ability to quickly and stably adapt to rapid changes in society, and even to rare and extreme events. From the viewpoint of social physics, this is fundamentally a question of how quickly social learning occurs: How can we most rapidly integrate data from everywhere, including unexpected and nontraditional sources, and then use these data to reliably reconfigure social systems?

Disaster management provides a familiar example of this type of system. In the face of unexpected devastation, and with only partially operating systems, how can we quickly restore basic functions? Our response to the Red Balloon Challenge system discussed in Chapter 7 showed the potential for social network incentives to guide rapid, distributed resource mobilization. Such examples give hope that we can build human-machine systems that very quickly configure both economic and social incentives to assemble entire systems, products, and services on the fly.

We need to think more broadly, however, than simply how to rebuild damaged systems. We also need to think about the resilience of the entire social design. Usually we think of finding the optimum strategy to manage a health care or transportation system, or to staff the system with the individuals who have the best training. But when there are systemic risks, such as hidden interdependencies or assumptions, then the entire system risks collapse. The poster children for such hidden dependencies are Lehman Brothers and AIG, whose collapse and near collapse demonstrated that most of the world's financial systems depend on largely unnoticed and unregulated financial activities.

Consequently, to survive systemic risks we need to have a diverse set of systems rather than one so-called best system. That way, should one of the systems fail, the surviving ones can quickly spread and take over from those failing. For example, an important lesson from Chapter 2 is that diversity is critical in decision systems, because although any particular strategy will always fail eventually, it is very *un*likely that many different types of strategy will fail at the same time. Similarly, one strategy for managing public health may fail completely under special conditions, just as the health system in New Orleans failed when telephone communications were wiped out during Hurricane Katrina. But the storm didn't destroy the amateur ham radio network, and so it stepped in to coordinate emergency delivery of medical drugs and equipment.

All of this suggests that in order to maintain the robustness of the entire society, we need a diverse set of competing social systems, each with its own way of doing things, together with fast methods of spreading them when required. This sort of robustness is exactly what we achieve when we tune a system for the best idea flow.

We are beginning to see such design principles in military and first-responder systems. They are being forced to acknowledge that not only can central decision-making capabilities be disrupted or unavailable, they can also be wrong, because they cannot judge local conditions as well as local commanders. As a consequence, these organizations are now beginning to train everyone in the system in the principles of distributed leadership. When decision making falls to those best situated to make the decision rather than those with the highest rank, the resulting organization is far more robust and resistant to disruption.

This is just a start, however. These hierarchical organizations have yet to admit the possibility that central commanders can also be wrong because those top leaders might simply have the wrong strategy. These kinds of organizations need to move beyond the big man theory of management and begin to build in better ways for continual testing of competing strategies.

Data for Development: D4D

Data about human behavior, such as census data, have always been essential for both government and industry. In this new era of big data, we must make sure that a digital data commons is freely available, and yet we must protect the privacy and safety of the individuals whose lives are reflected in it. Indeed, we need a New Deal on Data in which individuals can understand what information about them is used for and the benefits and risks of their use, so that they can choose how data will be shared both individually and collectively through government.

On May 1, 2013, we saw the public unveiling of what is perhaps the world's first true big data commons, with ninety research organizations from around the world reporting hundreds of results from their analysis of data describing the mobility and call patterns of the citizens of the entire African country of Ivory Coast.

These aggregated anonymous data were donated by the mobile carrier Orange, with help from the University of Louvain (Belgium) and my lab at MIT (United States), and in collaboration with Bouake University (Ivory Coast), the United Nations' Global Pulse, the World Economic Forum, and the GSMA (which is the mobile carriers' international trade association). The D4D initiative was led by Nicolas De Cordes (Orange), Vincent Blondel (Louvain), Alex Pentland (MIT), Robert Kirkpatrick (UN Global Pulse), and Bill Hoffman (World Economic Forum).

The ninety projects cover each of the three design criteria I have just proposed. An example of using the D4D data to improve social efficiency was highlighted by work done by researchers at the University College of London, who developed a method for mapping poverty from the diversity of cell phone usage.[19] This indirect method was first noticed by my former PhD student Nathan Eagle and relies on the wealth effect we saw in Chapter 9.[20] As people have more disposable income, their patterns of movement and phone calls become increasingly diverse. Another example of using the D4D data for social efficiency is the mapping of ethnic boundaries by researchers from the University of California, San Diego.[21] This method relies on the fact that ethnic and language groups communicate far more within their own group than with others. This project is significant because, while we know that ethnic violence often erupts along such boundaries, the government and aid agencies are usually uncertain about the geography of these social fault zones.

An example of using the D4D data for operational efficiency was an analysis by IBM's Dublin laboratory of Ivory Coast's public transportation system.[22] It showed that for very little cost the average commute time in Abidjan, the biggest city, could be cut by 10 percent. Other research groups demonstrated similar potentials for operational improvements in government, commerce, agriculture, and finance.

Finally, examples of using D4D data to improve resiliency include analyses of disease spread by groups from University of Novi Sad (Serbia), EPFL (Switzerland), and Birmingham (UK). They showed that small changes in the public health system could potentially cut the spread of flu by 20 percent as well as significantly reduce the spreads of HIV/AIDS and malaria.[23] These selected results are just a small sample of the impressive work that was made possible by this rich and unique data commons. These results and others like them are available at http://www.d4d.orange.com/home.

Each of these D4D research projects has demonstrated the great potential of a big data commons for improving society. From the point of view of Orange, it also demonstrates the potential for new lines of business that combine this data commons with individual personal data: Imagine a phone app that advises passengers about which bus will get them to work quickest, or how to reduce citizens' risks of catching the flu.

The work of these ninety research groups also suggests that many of the privacy fears associated with the release of data about human behavior may be generally misunderstood. In this data commons the data were processed by advanced computer algorithms (e.g., sophisticated sampling and use of aggregated indicators) so that it was unlikely that any individual could be reidentified. In fact, no path to reidentification was discovered by the several research groups that studied this specific question.

In addition, while the data were freely available for any legitimate research that a group was interested in, the data were distributed under a legal contract—similar to that used in trust networks—that specified that they could only be used for the purpose proposed and only by the specific people making the proposal. The use of both advanced computer algorithms and contract law to specify and audit how personal data may be used and shared is the goal of new privacy regulations in the EU, the United States, and elsewhere.

Summary: Promethean Fire

Throughout this book I have argued that we need to think about society as a network of individual interactions rather than as markets or classes. To accomplish this, I have presented a social physics framework that outlines how the flow of ideas from person to person shapes the norms, productivity, and creative output of our companies, cities, and societies.

By creating social systems that are based on using big data to map detailed patterns of idea flow, we can predict how social dynamics will influence financial and government decision making, and potentially achieve great improvements in our economic and legal systems. For one thing, we can begin using the tools of social physics to improve idea flow, and so hope to improve the productivity and creative output of our societies. Dense, continuous data together with visualizations of idea flow can also give us unprecedented instrumentation of how our policies are performing so that they can be quickly adjusted and revised as needed.

The first green shoots of this transformation are beginning to happen. All around the world governments and universities are beginning to take a new look at how cities are organized and governed, motivated by the rapid increase in city populations and the number of new cities that are being created. It is promising that many of these initiatives are reconsidering the basic design principles of cities and are beginning to take seriously the proposals of my colleagues and me to create a nervous system using mobile phone sensing and trust networks. At MIT the focus on research for city design is in full swing, and as codirector of MIT Media Lab's City Science initiative (see http://cities.media.mit.edu) I am now working with a variety of cities to improve their idea flow.

Basic to the vision of a data-driven society is the protection of personal privacy and freedom. To help guarantee such individual freedoms, I have been working with leading politicians, CEOs of multinational corporations, and public advocacy groups in the United States, the EU, and around the world to develop a New Deal on Data. These discussions have helped alter the privacy and data ownership standards around the world and are beginning to give individuals unprecedented control over data that are about them, while at the same time providing for increased transparency and engagement in both the public and private spheres.

We still face the challenge of more controlled experimentation within our social systems. The scientific method as currently practiced in the social sciences is failing us and threatens to collapse in an era of big data. One way forward is to construct living laboratories to test and prove out ideas for building data-driven societies.

In the end, I believe that the potential rewards of a data-driven society operating on the principles of social physics are worth the effort and the risk. Imagine: We could predict and mitigate financial crashes, detect and prevent infectious disease, use our natural resources more wisely, and encourage creativity to flourish and ghettos to diminish. These dreams used to be the stuff of science-fiction stories, but that fantasy could become a reality—our reality, if we navigate the pitfalls carefully. That is the promise of social physics and a data-driven society.

Appendix 1

Reality Mining

In recent years, the social sciences have been undergoing a digital revolution, heralded by the emerging field of computational social science. In our 2009 *Science* paper, David Lazer and I, together with more than a dozen endorsing colleagues, described the potential of computational social science to increase our knowledge of individuals, groups, and societies by use of data with an unprecedented breadth, depth, and scale.[1] The main driver of this revolution is the availability of big data about people and their behaviors: from credit cards, cell phones, Web searches, and more. The magazine *Technology Review* hailed the development of reality mining, the technology that enables much of this new computational social science, as one of the "10 technologies that will change the world."

My students and I have created two behavior-measurement platforms to speed the development of this new science. These platforms today produce vast amounts of quantitative data for hundreds of research groups around the world. The first platform is the sociometric badge, which replaces the humble ID badge with an electronic version that records the wearer's behavior; the second platform is behavior measurement software (funf) for our now ubiquitous smartphones. This appendix will briefly describe these two data-collection platforms.

The general framework for using both the sociometric badges and the smartphone funf system is a longitudinal living laboratory, or social-observatory type of study, coupled with a support-system infrastructure that enables the sensing and collecting of data, data processing, and a set of tools for feedback and communication with the subject population.

One of the key goals of these sorts of living lab experiments is to gather data simultaneously on numerous networking modalities (e.g., face-to-face, phone calls, e-mails, etc.) so that their properties and interrelations can be better understood. Typically we utilize the following components:

Digital sensing platform: This is the core of the study's data collection. Either a sociometric name badge or a smartphone is used as an in situ social sensor to map the users' activity features, proximity networks, and interaction patterns. Sociometric badges are best for experiments within companies where ID badges are already standard, whereas smartphones are best for living laboratory studies of entire communities.

Surveys: Subjects often complete surveys at regular intervals. Monthly surveys include questions about their self-perceptions, relationships, group affiliations, and interactions, as well as standard scales, such as the psychologists' Big Five personality test. Daily surveys might include questions about mood, sleep, and other activities, usually logged on the smartphone or a Web browser.

Purchasing behavior: Information on purchases is collected through receipts and credit card statements. This component targets categories that might be influenced by peers, such as entertainment and dining choices.

Digital social network data collection application: Participants can opt to install a social media application that logs information about their online social networks and communication activities.

When we compare the automatic digital measurements to the survey data we find surprising behavior patterns. For instance, just from data such as how much a subject walks around, who they call and when, and how much and when they socialize face-to-face, a user's personality type and disposable income can be estimated. We can also see when someone is coming down with the flu or is depressed.

Starting from these automatic digital measurements, it is easy to construct what social physics really cares about: networks and idea flow. These include face-to-face interactions, telephone calls, and social media networks. Almost as interesting, however, are location networks; i.e., who spends time in the same places. Or proximity networks; i.e., who goes to the same events.

Measurement of these networks gives us a picture of the subject's exposure to different ideas and experiences. From exposure, we can estimate the strength of the social influences between people and calculate the idea flow. This then enables the accurate prediction of group decision-making quality, productivity, and creative output.

Sociometric Badges

Figure 19: A standard design for a sociometric badge, courtesy of Sociometric Solutions, Inc.

The most valuable flows of ideas within an organization are face-to-face and telephone conversations, because they carry the most complex, sensitive information—yet few organizations measure them. And, of course, what isn't measured can't be managed.

Our research has included innovation teams, post-op wards in hospitals, customer-facing teams in banks, backroom operations, and call center teams. We typically equip all members of those organizations (especially the managers) with sociometric badges (such as the one shown in Figure 19) that collect data on their individual communication behaviors—tone of voice, body language, whom they talked to and how much, and more. With remarkable consistency, we have found patterns of communication to be the most important predictor of a team's success. Not only that, but they are often as significant as all the other factors—individual intelligence, personality, skill, and the substance of their ideas—combined.

The sociometric badge in Figure 19 collects and analyzes social behavior data by measuring many of the common social signals expressed by the wearer. It incorporates a location sensor, accelerometers to record body language, a proximity sensor to determine who else is around, and a microphone that notes whether anyone is speaking. To avoid privacy violations, however, the device does not record speech content or video.

It is designed to be worn around the neck, like a typical company ID badge. They are put on when the subject comes into work and taken off when they leave. The one difference in use is that they have to be plugged into a USB port of a computer or a charging station in order to recharge the batteries.

The capabilities of the sociometric badge include:

- extracting measurements of energy, engagement, and exploration by pooling the measurements of multiple participants;
- measuring personal energy levels, the amount of extraversion and empathy shown in body language, and the rhythmic patterns associated with flow states.

The badges can be used to give teams real-time feedback on group interaction patterns, which is particularly useful for virtual or distance teams. They are now manufactured by the MIT spin-off Sociometric Solutions (which I cofounded) for use in its business consulting practice and are available for research on a not-for-profit basis.

The data collected by the badges are changing how people lay out office space and altering how companies understand patterns of interactions. These sorts of data are particularly important for long-distance work and cross-cultural teams, which are so crucial in a global economy, because they can now visualize and improve their interaction patterns.

As of 2013, many dozens of research groups have begun using

the sociometric badge for social physics research. In addition, dozens of companies, including many Fortune 1000 companies, are now using them for space and reorganization planning. For additional information, see www.sociometricsolutions.com.

Mobile Phone Sensing

My students and I utilized smartphones and pervasive computing methodologies to develop a mobile phone-centric social- and behavioral-activity sensing system called funf. The data collected by funf include continuous collection of over twenty-five phone-based signals—including location, accelerometry, Bluetooth-based device proximity, communication activities, installed applications,

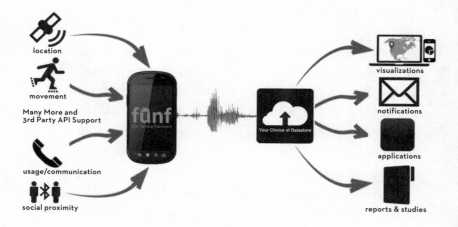

Figure 20: funf mobile phone sensing system.

currently running applications, and multimedia and file system information, as well as additional data generated by our experimental applications. In addition, we collect financial information through receipts and credit card statements; logs of digital social media activities; daily polling of moods, stresses, sleep, productivity, and socialization; other health- and wellness-related information; standard psychological scales, such as personality tests; and many other types of data manually entered by the participants.

These data enable us to automatically reconstruct multiple network modalities of the subject's community, such as their phone communications, physical face-to-face encounters, and online social relationships, along with manually self-reported networks. We use these network observations to investigate how things such as ideas, decisions, mood, or the seasonal flu are spread in the community. Our high-level goals include: the investigation of "natural" and externally imposed social mechanisms related to behavior and decision making, and designing and evaluating new mechanisms or tools for helping people make better decisions.

The funf Open Sensing Framework, described in Ahrony et al. 2011, is an extensible sensing and data-processing framework for mobile devices. It provides an open-source, reusable set of functionalities that enable the collection, uploading, and configuration of a wide range of data types. Today more than fifteen hundred groups around the world are using funf.

It is intended for scientific research, and consequently, one of the central concerns when using funf is the protection of privacy and sensitive information. Thus, all of funf's functions include strict privacy protection measures. For example, data are linked to identifiers coded for phone users and not to their real-world personal identifiers. All text that is readable by humans, such as phone numbers and text messages, is captured as hashed identifiers and never saved in clear text.

Examples of standard funf sensing functions are:

GPS
WLAN
Accelerometer
Bluetooth
Cell tower ID
Call log
SMS log
Browser history
Contacts
Running apps
Installed apps
Screen state
Media battery status

Social media activity, credit card activity, and other sorts of information can also be recorded. It is available for Android mobile phones at http://www.funf.org.

Appendix 2

OpenPDS

Personal data—digital information about users' locations, calls, Web searches, and preferences—have been called the oil of the new economy and what I have seen reinforces this comparison.[1] These high-dimensional data are what allow apps to provide smart services and personalized experiences. From Google's search to Netflix's "movies you should really watch," from Pandora to Amazon, data provide the fuel that powers these and hundreds of other services. The algorithms help users become more connected, productive, and entertained. These applications are also emblematic of both the amazing potential and possible risks associated with user-centric data.

Personal and user-centric data are already collected, processed, and leveraged on a large scale. They are collected and stored by hundreds of different services and companies. Such fragmentation makes the data inaccessible to innovative services, and often to the individual who generated it in the first place. This prevents users from taking full advantage of their data and makes it very hard, if not impossible, for an individual to understand and manage the associated risks. Because most of the data are not anonymous, or could be reidentified, this is a major concern. Advancements in using and mining these data must evolve in parallel with considerations about ownership and privacy.

Toward Personal Data Stores

Data ownership and repositories of personal data have been discussed for a long time. The large-scale deployment of such solutions is a chicken-and-egg problem, however, because users are waiting for compatible services while services are waiting for user adoption.

Recent political and legal developments are a game changer in this dilemma, as my work with John Clippinger at the Institute for Data Driven Design has shown.[2] The framework that PhD and postdoctoral students Yves-Alexandre de Montjoye, Erez Shmueli, Samuel S. Wang, and I have developed, called openPDS,[3] uses the World Economic Forum definition of "ownership" of data that I proposed as the New Deal on Data, i.e., the rights of possession, use, and disposal.[4] In addition, it follows the policies of the National Strategy for Trusted Identities in Cyberspace (NSTIC),[5] the U.S. Department of Commerce green paper, and the U.S. International Strategy for Cyberspace.[6] The framework, openPDS, is also strongly aligned with the European Commission's 2012 reform of the data protection rules.[7] These recommendations, proposed reforms, and regulations all recognize the increasing need for personal data to be under the control of the individual, as he is the one who can best judge the balance between the associated risks and rewards.

At a time when users are interacting with numerous companies on a daily basis, interoperability is not enough to achieve practical data ownership, let alone address privacy concerns. To achieve true data ownership, users need to own a secure space acting as a centralized location where their data can live. Owning a personal data store (PDS) would allow the user to view and understand how the data collected might be used, as well as to control the flow of data and to manage fine-grained data access.

In addition to facilitating data ownership, a PDS is also a particularly attractive solution because it enables a fair and efficient market for data, i.e., a market where users can get the best services and algorithms for their data.[8]

> *Fair:* Users are the ones controlling access to their data so services can be rated and evaluated. Users can decide whether a service provides enough value for the amount of data it requests, taking into account the company's reputation. In the proposed framework, a user would be empowered to ask questions such as: "Is finding out the name of this song worth enough to me to give away my location?" and can otherwise easily switch his data to another service.

> *Efficient:* Users can seamlessly interface and give new services access to their data. The proposed framework removes barriers to entry for new businesses, allowing the most innovative companies to provide better data-powered services. It also incentivizes businesses, as the services chosen by the user can avoid collecting most of the data themselves. Businesses will further gain access to historical data that have been collected by sensors on the smartphone and/or by other apps and services. Service providers can thus concentrate on delivering

the best possible experience to the user using all the available data. For example, a music service can provide personalized radio stations by leveraging the songs and artists users said they like across the Web, what their friends like, or even which nightclubs they visit.

Other approaches have been proposed for the storage, access control, and privacy of personal data. OpenPDS is unique, though, in both its alignment with current political and legal thinking and its dynamic privacy-preserving mechanism.

Protecting the privacy of personal data is known to be a hard problem. The risks associated with high-dimensional data are often subtle and hard to predict.[9,10] Anonymizing individual unaggregated data is a challenge that experts have called "algorithmically impossible."[11] Over the last few years, numerous efforts have exposed the risks of reidentification or deanonymization of seemingly anonymous data sets. For example, mobility data sets of millions of users were found to be potentially reidentifiable using only four spatiotemporal points.[12]

Dynamic Privacy: A New Paradigm

Numerous ways of protecting or obfuscating personal data have been proposed; however, none of them has proven satisfactory for the high-dimensional, multimodal, continuously evolving data that is being recorded today. Instead we have developed the idea of dynamic privacy in order to turn an algorithmically impossible anonymization problem into a more tractable security one by answering questions instead of giving access to the raw data.

Imagine a service that aims at personalizing a user's experience based on whether he is currently running or not. Under the existing model, the service would collect the location and/or accelerometer data from the user's mobile phone and upload it to a remote server in order to compute the relevant piece of information— running or not running. Under the openPDS/dynamic privacy mechanism, a piece of code would be installed inside the user's PDS. The installed code would use the sensitive location and accelerometer data to compute the relevant answer within the safe environment of the PDS. That answer alone would be sent to the remote server.

Combined with data ownership, this simple idea allows users to benefit from a personalized experience without having to share raw data, such as raw accelerometer readings or GPS coordinates. In other words, the code is shared, not the data. While this is not a complete solution by itself, on-the-fly reduction of the dimensionality and scope of the data so that only the minimum needed for the specific problem is shared makes sharing safer. Such a mechanism also allows users to safely grant and revoke data access, to share data anonymously without needing a trusted third party, and to monitor and audit data uses. A group computation mechanism takes this even one step further by enabling users to anonymously contribute data for use in an aggregate form to answer questions such as: "How many users are currently in this region?"

User experience: Suppose Alice wanted to install and use an Android application such as Foursquare, the location-based check-in system, without using a PDS. Alice downloads the application onto her phone and authorizes Foursquare's access to her phone's network communications, personal information, and phone features decisions; users already face this problem when installing any new application on an Android phone. Alice would create a user account, and then start building a relationship with Foursquare from scratch.

Foursquare would store all the information it collects about Alice in its back-end servers. Alice would have no ability to access that data or see what Foursquare uses to make inferences about her over time. In addition, any integration between different services happens behind the scenes. If Foursquare wished to leverage Twitter or Facebook data, Alice would have to authenticate to those services, but the amount of outside data that Foursquare then uses is mostly unknown to Alice.

If Alice chooses to download a PDS-aware version of Foursquare, she would install it just as she would any other Android application. Upon launching it, the Android app would prompt her to install a Foursquare app onto her PDS. The PDS app would describe exactly what data Foursquare would access on her PDS, as well as what relevant summarized information is passed on to Foursquare's servers, allowing Alice to understand what it means for her privacy to install the app.

Rather than storing Alice's personal data on Foursquare's servers, the Foursquare PDS app would instead access and process the data on Alice's PDS. Alice would have installed or bought a PDS on her favorite cloud provider, or on her own server. Over time her PDS would be filled with information collected by her phone, along with information about her musical tastes, her contacts, and a stream of other sensor information that she accumulates in her day-to-day life. Alice would have full control over these data and could see exactly what her phone, other sensors, and services gather about her over time.

Because the Foursquare PDS app is being run on a computing infrastructure that Alice owns, the outgoing data can be audited to verify that no unexpected data are escaping the boundaries of her PDS. In this way, rich applications and services can be built on top of the PDS that leverage all of these disparate data sources, while Alice still owns the underlying data behind these computations and can take steps to preserve aspects of her privacy.

An example application: Mental diseases rank among the top health problems worldwide in their cost to society, even though they are often quite treatable. Major depression, for instance, is the leading cause of disability in established market economies. Diagnoses of psychiatric disorders are overwhelmingly based on reporting by the patient, a teacher, family member, or neighbor.

Many symptoms of psychiatric disorders concern patterns of physical movement, activity, and communication—all things that can be measured by mobile phone data. Accelerometers can reveal fidgeting, pacing, and abrupt or frenetic motions. Location tracking can reveal changes in places visited and routes taken as well as the overall extent of physical mobility. The frequency and pattern of individuals' communications with others and the content and manner of their speech can also reflect key signs of several psychiatric disorders. Moreover, the value of these data are multiplied when combined with an occasional question about how the person is feeling or what they are doing at the moment their behavior begins to become worrying.

If we could passively and automatically measure these "honest signals" of mental distress, then medical personnel could potentially reach people before their lives spiraled out of control. Even more important, if their friends could receive hints that something was going wrong, then they could reach out at just the point when support from friends has an enormous power to help. There is, of course, the problem of privacy.

My observations about how mental health could be assessed by the honest signals of behavior that can be measured by the sensors in mobile phones and about the value of users sharing these signals with their friends inspired the inclusion of the openPDS and funf systems into DARPA's Detection and Computational Analysis of Psychological Signals program (DCAPS).[13]

In DCAPS, smartphones provide a pervasive platform that enables continuous sensing and monitoring in natural settings while minimizing the effort burden placed on veterans. These devices can record the user's tone of voice, frequency of interactions with other people, general levels of movement and activity, as well as other subtle and honest social signals. In fact, a sizable proportion of the current DSM-IV symptoms that are used to diagnose various psychological health conditions focus on changes in behavior that are precisely the types of measurements that can be effectively captured by smartphone interactions (DSM-IV, soon to be updated as DSM-V, is the most widely accepted mental health diagnostic manual).

Our openPDS and funf systems provide a privacy-assured, secure, and scalable mobile sensing platform that gathers honest signals data from mobile phones, which can then be analyzed for patterns of psychological distress. The data can be stored securely by openPDS, and it enables each individual to see and share feedback on their overall mental health status.

Figure 21 shows the DCAPS mobile phone interface developed in my lab at MIT (other DCAPS contractors developed their own interfaces). In this display, people can see three dimensions of how their behavior was rated during the previous day. These dimensions include activity level, amount of socialization, and level of focus during activities. All three of these dimensions are DSM-IV criteria for depression and post-traumatic stress disorder (PTSD) but are also commonsense dimensions of day-to-day life, and so their meaning is understandable both by the user and medical personnel. In Figure 21, the

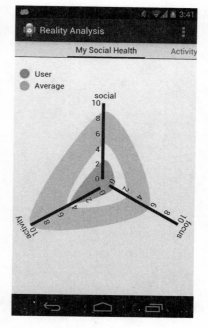

Figure 21: Secure and private measurement of mental health on mobile phones.

central cluster shows the user's behavior over the previous day, surrounded by a ring that shows the maximum and minimum values of activity, focus, and socialization for the user's buddies. In this case, the user is pretty clearly an outlier relative to his peers, and this should encourage both the user and the buddies to think about discussing the cause of this situation.

Appendix 3

Fast, Slow, and Free Will

Psychologist Daniel Kahneman and artificial intelligence pioneer Herb Simon, both Nobel Prize winners, each embraced a model of a human mind with two ways of thinking.[1] In Kahneman's formulation, one way of thinking is a fast, automatic, and largely unconscious mode, and the second way of thinking is a slow, rule-based, and largely conscious mode. A thumbnail sketch of fast thinking is that it drives habits and intuitions, largely by using associations among personal experiences and the experiences learned by observing others. In contrast, the slow mode of thinking uses reasoning, combining beliefs in order to reach new conclusions.

Fast thinking: Fast thinking is an older system and great for getting answers instantly, even those requiring complex trade-offs, and it is also adept at finding patterns and associations. Our fast-thinking capability excels at learning through our own experiences and through exposure to the experiences of others; however it is limited to forming associations rather than employing abstract reasoning. Fast thinking is something that we inherited from our ape ancestors, and it is likely that the mental capabilities of early humans were based largely on this system.

Fast thinking is built on taking ideas—an action to take, the context in which to take it, and the likely outcome—that seem useful and making them models for future behavior. Because fast thinking is so automatic and unconscious, it pays to be very conservative about which ideas are used as a basis for it. It is no surprise, therefore, to find that we are slow to learn new behavior habits. It often takes many examples of an idea being successful before we will build that idea into our suite of habits, and thus engagement with others who are experimenting with the same ideas is the typical way that we learn new habits.[2] Observing the experiences of others provides us with the examples we need to decide whether or not a new idea will be successful for ourselves.

Slow thinking: While fast thinking works very well—its roots seem to be at least a hundred million years old and may be shared by all mammals—it also has some real drawbacks, due to inherent limitations in its use of association as the mechanism for choosing the right action for the current situation. In fact, Kahneman and others have speculated that these limitations probably spurred the evolution of slow thinking.

Slow thinking is built on beliefs gained by individual reasoning and observations that seem interesting—facts that might prove useful someday. Because slow thinking is rule-based and reflective, it is safe to entertain new and uncertain beliefs, because by "playing" with them, we will eventually determine if they fit with everything else we believe. It makes sense, then, that we are quick to learn new facts and that we continually engage in exploratory behavior. As we saw in Chapter 2, exploratory behavior improves our ability to make good decisions.

Language and slow thinking are tightly coupled. While we sometimes integrate the experiences of others into our fast-thinking repertoire of habits through particularly memorable stories, the real power of language is that it allows the belief structures of slow thinking to be spread through a population. The ability to go beyond the bounds of here-and-now, familiar experiences may be the key contribution that slow thinking makes to the fitness of our species, even though it is usually quite a slow and effortful way to arrive at answers.[3]

It surprises most people to learn that fast thinking is better than slow thinking for many tasks.[4] Whenever a problem is complex and involves trade-offs between different goals, the association mechanisms used in fast thinking usually outperform the slower reasoning mechanisms. This is especially true when there is limited time to make a decision. For this reason, many scientists believe that the vast majority of our daily behavior is due to fast thinking—we literally don't have the time to think things through using slow thinking.[5] The power of fast thinking is most visible in emergencies, when people often say "I didn't even think about it, I just responded." The same logic applies equally well to our humdrum daily routines, during which people are often daydreaming or chatting away while at the same time filing papers or driving a car.

So while we may make a high-level, conscious decision to engage in some activity, many of the activities themselves are highly
practiced and automatic, driven by fast thinking and largely out of
the spotlight of our attention. The largely automatic nature of our
lives is most visible when we are expert at an activity, such as performing the daily routines of life, engaging in social chitchat, or
physical activities like driving or riding a bike. For these habitual
activities, we are typically hard-pressed to explain exactly what we
did or why we did it, because we were simply on autopilot.

Fast and slow together: It is often hard to pull out the details of
the interactions between fast and slow thinking, because evolution has bound them together so tightly. I think that what we are
seeing in the Social Evolution and Friends and Family studies is
mostly the role of the fast habitual or intuitive mode of thinking,
which learns new behaviors in a similar manner across people and
situations. In contrast, our slow-reasoning mode of thinking is too
diverse and complex to show up as a single factor in these experiments, except when it comes to choosing which idea stream to
join. We can begin to investigate, however, how fast and slow
thinking work together through big data experiments such as were
reported on in Chapter 3.

My studies suggest that the continual exploratory behavior of
humans is typically a conscious, slow-thinking process and that it
is guided by social exposure through all the different communication channels, e.g., by apparent popularity among peers. This sort
of peer pressure is informative, not normative.

While broad social exposure guides the slow-thinking process
of exploration, it is not closely connected to the learning process for

fast thinking. Whether or not some newly discovered behavior is integrated into the fast-thinking repertoire of habits and intuitions has more to do with how the particular behavior fits into the long-term common sense developed with a community of peers.

The best capsule summary is that habits and gut instinct are based on fast thinking, which uses engagement with others to integrate their experiences with our own, and thus form our habits of action. Exploration and guiding our attention to help figure things out seem to be the core functions of slow thinking, which is supported by observations of events, context, and correlation that are learned through both personal perception and language.[6]

Understanding that humans have two systems of thinking that work quite differently transforms many of the classic disputes in philosophy, anthropology, and sociology.[7] On one side of this academic battle are anthropologists such as Claude Lévi-Strauss, philosopher-economists such as Karl Marx and Adam Smith, and many social psychologists. Thinkers on this side of the dispute emphasize how the structure of society shapes the behavior of the individual. On the other side of the battle are philosophers such as Jean-Paul Sartre, game theorists, and cognitive scientists, who emphasize free will and how individual cognitive processes shape individual behavior.

The modern discovery that the human mind has two types of thinking yields this conclusion: It tells us that both sides of the free will versus social context debate are right, but that neither is right about all of human behavior all of the time. For instance, we found in the Social Evolution study of political beliefs that people apparently used the tools of slow thinking to decide whether they felt more comfortable around liberals or conservatives. But after they made that choice, the automatic-learning tools of fast thinking caused them to absorb the relevant intuitions and habits of their chosen group.

From a quantitative point of view, however, the social influence side wins. The vast majority of our behavior is habitual rather than reasoned, which runs counter to how many of us would like to view ourselves.[8] As Kahneman put it, most of our behavior is based on the fast judgments of intuition and habit, not the slow process of reasoning. But, as the free will side would point out, it is likely that the majority of our most important decisions are due to the slow process of reasoning.

Appendix 4

Math

How can we model influence, social learning, and peer pressure between individuals in a social system even when the network of interactions is unknown? In this appendix, I will: 1) review the influence model, which utilizes independent time series to estimate how much the state of one actor affects the state of another actor in the system; 2) show how this general architecture can be used to model social learning across several modalities; 3) demonstrate how to predict the propagation of behavior change (idea flow) across a social network; and 4) explain how to use social network incentives to alter this flow.

The concept of influence is extraordinarily important in the natural sciences. The basic idea of influence is that an outcome in one entity can cause an outcome in another. Flip over the first domino, and the second one will fall. If we understand exactly how two dominoes interact—how one domino influences another—and we know the initial state of the dominoes and how they are situated relative to one another, then we can predict the outcome of the whole system.

For decades, social scientists have also been interested in analyzing and understanding who influences whom in social systems. But the analogue with the physical world is behavior specific, and the situational context of specific interactions can change the effect one actor has on another. And even more challenging, actors can choose with whom they interact, which can confound efforts to infer influence from correlated behaviors between actors. As a result, there has been tremendous interest in developing methods for better understanding the effects that networked interactions have on the spread of social behaviors and outcomes.

Social scientists have already carefully studied communication settings such as group discussions to better understand the causal mechanisms that underlie influence, but recent advances in modern sensing systems, such as sociometric badges and cell phones, now provide valuable social behavioral signals from each individual at very high resolution in time and space. The challenge is how to use these data to make better inferences about influence within social systems.

In this appendix, I begin by describing the influence model whose current version was first developed by Wen Dong and me in his master's thesis, and later refined by Wei Pan, Manuel Cebrian, and Taemie Kim in my lab, together with social scientist James Fowler of the University of California, San Diego.[1] This appendix uses the formulation described in Pan et al.[2] Similar definitions of influence in other literature include research on voting models in physics, cascade models in epidemiology, attitude influence in psychology, and information exchange models in economics.

Previous models of influence, however, have been difficult or impossible to use to predict behavior change from real-world observations. Classic diffusion models such as Granovetter's work are applicable to simulation but lack data-fitting and prediction powers.[3] Statistical analyses used by social scientists, such as matched sample estimation, are useful only for identifying network effects and mechanisms.[4] Recent works in computer science for inferring network structure assume a simple diffusion mechanism and are only applicable to artificial simulation data on real networks.[5]

Our work addresses each of the above issues, as demonstrated in the several examples reported in this book. Our influence model is not, of course, the only way to model the dynamics of social interaction. It is not *the* model of social physics. It is, however, a scalable, efficient method that can handle heterogeneous individuals, changing social relationships, and missing or noisy data, and one that has proven itself adequate to the task in many, many situations.

The influence model is built on an explicit abstract definition of influence: An entity's state is affected by its network neighbors' states and changes accordingly. Each entity in the network has a specifically defined strength of influence over every other entity and, equivalently, each relationship can be weighted according to this strength.

Our experimental results in the Friends and Family study, the Social Evolution study, and elsewhere have shown that the amount of exposure to peers who have already adopted a particular behavior can provide a good estimate of the probability that an individual will adopt that behavior, at least for the behaviors where the actions and outcomes are visible. This is why social physics works. Without these sorts of strong social learning and social pressure effects we would instead have to model the detailed thought patterns of each individual.

Thus, by combining the influence model with a measure of exposure to peers' behaviors, or a measure of social tie strength, we can produce useful predictions of the likelihood that an individual will adopt a particular behavior. Typical results are that 40 percent of the variance in behavior adoption can be predicted in this manner; that is, our influence predictions are roughly as powerful a predictor of behavioral outcomes as IQ or genetic makeup.

I believe that the influence model is a unique tool for social scientists because it can be applied to a wide range of social systems (including those where aggregates such as organizations, states, and institutions can themselves be thought of as actors in a network). The influence model also enables researchers to infer interactions and dynamics when the network structure is unknown—all that is needed is information about time-series signals from individual observations.

Although this modeling method is subject to the same limitations as any observational network study, the ordering of behaviors in time and social space makes it less likely that alternative mechanisms, such as selection effects and contextual heterogeneity, can explain the patterns of influence ascertained by the model.

Influence Between Entities

The influence model my students and I developed to model human social interactions begins with a system of entities C. Each entity c is an independent actor, which can be a person in the case of, say, a group discussion. These entities interact and influence each other along a social network. Influence is defined as the conditional dependence between each entity's current state $h_t^{(c)}$ at time t and the previous states of all entities $h_{t-1}^{(c)}$ at time $t-1$. Therefore, intuitively, $h_t^{(c)}$ is influenced by all other entities. An important implication of this Markovian assumption is that all effects from states at times earlier than $t-1$ are completely accounted for by incorporating all information from time $t-1$. This does not mean that earlier times had no effect or were unimportant; it just means that their total effect is felt in the immediately previous time period.

Each entity c is associated with a finite set of possible states $1, \ldots, S$. At a time t, each entity c is in one of these states, denoted by $h_t^{(c)} \in (1, \ldots, S)$. It is not necessary that each entity be associated with the same set of possible states, but for simplicity, assume that each entity's latent state space is the same without loss of generality. The state of each entity is not directly observable. As in a hidden Markov model (HMM), however, each entity emits a signal $O_t^{(c)}$ at time stamp t based on the current latent state $h_t^{(c)}$, following a conditional emission probability $\text{Prob}(O_t^{(c)}|h_t^{(c)})$.

Defining social influence in terms of state dependence—how an entity's state impacts other entities' states and vice versa—is an idea rooted in statistical physics and machine learning. Similarly, researchers have long used Bayesian networks to understand and process social interaction time-series data. Earlier projects used coupled HMMs, while more recent ones have employed dynamic system trees and interacting Markov chains. Our model is unique, because it connects social networks to state dependence parsimoniously.

A social system comprises many entities interacting with and influencing one another. Social influence can be expressed as the conditional dependence between each entity's current state $h_t^{(c)}$ at time t and the previous states of all entities $h_{t-1}^{(1)}, \ldots, h_{t-1}^{(C)}$ at time $t-1$. Therefore, intuitively, $h_t^{(c)}$ is influenced by all other entities. The conditional probability is written

$$\text{Prob}(h_t^{(c')}|h_{t-1}^{(1)}, \ldots, h_{t-1}^{(C)}) \tag{1}$$

and naturally describes a generative stochastic process. Using a coupled Markov model, we can take a general combinatorial approach and convert (1) into an equivalent HMM, in which a unique state represents each different latent state combination $h_{t-1}^{(1)}, \ldots,$ $h_{t-1}^{(C)}$. Therefore, for a system with C interacting entities, the equivalent HMM will have a latent state space of size S^C, exponential to the number of entities in the system, which is unacceptable in real applications.

In contrast, the influence model uses a simpler approach with dramatically fewer parameters. Entities $1, \ldots, C$ influence c' in the following way:

$$\text{Prob}(h_t^{(c')}|h_{t-1}^{(1)}, \ldots, h_{t-1}^{(C)})$$
$$= \Sigma_{c = (1, \ldots, C)} \, R_{c',c} \times \text{Prob}(h_t^{(c')}| h_{t-1}^{(C)}) \tag{2}$$

R is a $C \times C$ row stochastic matrix that models the tie strength between entities. $\text{Prob}(h_t^{(c')}| h_{t-1}^{(C)})$ is modeled using an $S \times S$ row stochastic matrix $M^{c,c'}$, which describes the conditional probability between states of different entities and is known as the transition matrix. Generally, for each entity c there are C different transition matrices to capture the influence dynamics between c and $c' = 1, \ldots, C$. However, this situation can be simplified by replacing the C different matrices with only two $S \times S$ matrices, E^c and F^c: $E^c = M^{c,c}$ captures the self-transitions, and because entity c's influence over other entities is similarly fixed, the interentity state transitions $M^{c,c'} = F^c$ for all $c' \neq c$.

Equation 2 can be viewed as follows: All entities' states at time $t{-}1$ will influence the state of entity c' at time t. However, the influence strength varies for different entities. $R^{c',c}$ captures the strength of c over c'. The state distribution for entity c' at time t is a combination of influence from all other entities weighted by their strength over c'. Because R captures influence strength between any two entities, we refer to R as the "influence matrix."

The number of parameters grows quadratically with respect to the number of entities C and the latent space size S. This largely relieves the requirements for large training sets and reduces the chances of model overfitting, making the influence model scalable to larger social systems. In addition, R can be naturally treated as the adjacency matrix for a directed weighted graph. The influence strength between two nodes learned by the model can then be treated as a tie weight in social networks. In this way, the model connects conditional probabilistic dependence to a weighted network topology. In fact, the most common usage for R is to understand social structure. Matlab code for estimating influence model parameters and example problems are available at http://vismod.media.mit.edu/vismod/demos/influence-model/index.html.

The influence model has been applied to various social science experiments, particularly those that have been monitored by sociometric badges or smartphones.[6] These include investigating conversational turn taking, dominance in social networks, and understanding human interaction contexts. For instance, my students and I have used the influence model to understand the functional role (such as follower, orienteer, giver, and seeker) of each individual in the mission survival group discussion data set.[7] We found that the inferred influence matrix helped them to achieve better classification accuracy compared with other more traditional approaches. Recently, the influence model has been extended to a variety of systems, including traffic patterns[8] and flu outbreaks.[9] In addition, there have been methodological advances that allow the model to incorporate dynamic changes in the influence matrix itself.[10]

Related approaches have utilized Bayesian networks to understand and process social-interaction time series data. Examples include coupled HMM, dynamic system trees, and interacting Markov chains. The key difference between these approaches and the influence model is that the influence matrix R connects the real network to state dependence.

The Inverse Problem: Inference of latent variables. In most practical situations we are given only an observation time series that consists of the measurements of behavior. Based on these observations, we need to learn the distributions of underlying latent variables and system parameters for the influence model. In our work we use a variational expectation maximization (EM) approach, although one could also use a mean-field method. Please refer to Pan et al. for detail.[11]

Discussion: I have described the influence model and how it can be applied to a variety of social signals to infer how entities in networks affect one other. In particular, we can use the resulting influence matrix R to connect the underlying social network and the stochastic process of behavior state transition of the individuals being observed.

The influence model shares some of the same limitations as

other machine-learning models: Inference requires sufficient training data and tuning is necessary to get the best results. The most important limitation is that we are attempting to infer causal processes from observational data in which many mechanisms are likely at play. If we find that behaviors between two individuals are correlated, it could be due to influence, but it could also be due to selection (I choose to interact with people like me) or to contextual factors (you and I are both influenced by an event or by a third party not in the data). Recently it has been shown that these mechanisms are generically confounded. The fact that we have time data to test causal ordering as well as asymmetries in network relationships to test the direction of effects, however, means that we can have greater (but not complete) confidence than we would if we only had cross-sectional data from symmetric relationships.

Modeling Influence Through Multiple Channels (Chapter 3)

Modern smartphones, such as the ones used in many of my studies, are able to capture many different sorts of social networks using their built-in sensors. These include call contacts lists, people who are often nearby, and people who share the same mobility habits. Each of these networks exposes an individual to new ideas, and hence offers opportunities for social learning.

Our experimental results in the Friends and Family study, the Social Evolution study, and elsewhere have shown that the amount of exposure to peers who have already adopted a particular behavior can be used to generate a good estimate of the probability that an individual will adopt that behavior, at least for the behaviors where the actions and outcomes are visible. The next question, though, is how to extend the influence model so that it operates with different influence parameters within each modality in order to predict behavior change resulting from multiple channels of exposure.

PhD student Wei Pan, working with me and PhD student Nadav Aharony, developed a simple computational model to better predict behavior change: using a composite network computed from the different networks sensed by phones. Our model also captures individual variance and exogenous factors in behavior change. We show the importance of considering all these factors in predicting behavior change and, finally, observe that behavior change is indeed predictable. The formulation here is from Pan et al.[12]

Introduction: My recent research projects have demonstrated that social network exposure correlates with individual behavior changes, such as weight gain, voting behavior, etc.[13] Here we are interested in extending the influence model's capability for network-based prediction, in order to address the challenge of utilizing the many different types of network data that have been obtained from sensors such as smartphones for the purpose of obtaining more accurate and more general behavior change prediction.

It is difficult to adopt existing tools from large-scale social network research in order to model and predict behavior change due to the following facts:

1. The underlying network is not fully observable. In this work, our key idea is to infer an optimal composite network, i.e., the

network that best predicts behavior change, from multiple layers of different networks easily observed by sensors such as modern smartphones, rather than assuming that a certain network is the "real" social network for explaining behavior change.

2. There are exogenous factors in behavior changes. Network analysis for behavior change often assumes that transmission along observed networks is the only mechanism for adoption. This is, of course, simply not true; there are mass media and unobserved networks that may cause behavior change. One major contribution of our work in this area is that we demonstrate that it is still possible to build a useful prediction tool despite such randomness.

3. The individual behavioral variance in behavior change could be so significant that any network effect might possibly be rendered unobservable from the data. For instance, some people are early adopters and others are late adopters.

In this section, I describe our model for capturing the behavior changes in networks. In the following, G denotes the adjacency matrix for graph G. Each user is denoted by $u \in \{1, \ldots, U\}$. Each behavior is denoted by $\mathbf{a} \in \{1, \ldots, A\}$. We define the binary random variable $\mathbf{x}_u^a$ to represent the status of adoption (e.g., app installation): $\mathbf{x}_u^a = 1$ if $\mathbf{a}$ is adopted by user u and 0 if not. As introduced in the previous section, the different social relationship networks that can be inferred by phones are denoted by $G^1, \ldots, G^M$. Our model aims to infer an optimal composite network G^{opt}, with the most predictive power from all the candidate social networks. The weight of edge $e_{i,j}$ in graph G^m

is denoted by $w_{i,i}^m$. The weight of an edge in G^{opt} is simply denoted by $w_{i,j}$.

One basic idea of our model is the nonnegative accumulative assumption, which distinguishes our model from other linear mixture models. We define G^{opt} to be:

$$G^{opt} = \sum_m \alpha_m G^m$$

where $\forall m,\ \alpha_m \geq 0$.

The intuition behind this nonnegative accumulative assumption is as follows: If two nodes are connected by a certain type of network, their behaviors may or may not correlate with each other; on the other hand, if two nodes are not connected by a certain type of network, the absence of the link between them should lead to neither positive nor negative effect on the correlation between their app installations. Thus, $\alpha_1, \ldots, \alpha_M$ are the nonnegative weights for each candidate network in describing the optimal composite network. We continue to define the network potential $p_a(i)$:

$$p_a(i) = \sum_{j \in N(i)} w_{i,j} x_j^a$$

where the neighbor of node i is defined by

$$N(i) = \{j \mid \exists m \text{ s.t. } w_{i,j}^m \geq 0\}$$

The potential $p_a(i)$ can also be decomposed into potentials from different networks. We can think of $p_a(i)$ as the potential of i showing a new behavior based on the observations of its neighbors on the composite network. Finally, our conditional probability is defined as:

$$Prob(x_u^a = 1 \mid x_{u'}^a : u' \in N(u)) = 1 - exp(-s_u - p_a(u))$$

where $\forall u$, $s_u \geq 0$. s_u captures the individual susceptibility to behavior change. We use the exponential function for two reasons:

1. The monotonic and concave properties of $f(x) = 1 - exp(-x)$ match with recent research on human behavior change through social influence, which suggests that the probability of adoption increases at a decreasing rate with increasing external network signals.[14]
2. It forms a concave optimization problem during maximum likelihood estimation in model training.

We must still account for exogenous factors such as the popularity of a behavior. We can model this by introducing a virtual graph G^p which can be easily plugged into our composite network framework. G^p is constructed by adding a virtual node $u + 1$ and one edge $e_{u+1,u}$ for each actual user u. The corresponding weight of each edge $w_{u+1,u}$ is a positive number describing the popularity of the behavior.

Inclusion of these exogenous factors also increases accuracy in measuring network effects for a nontrivial reason. For example, consider a network of two nodes connected by one edge, and both nodes exhibit a behavior. If this behavior is very popular, then the fact that both nodes have this behavior may not imply a strong network effect. On the contrary, if this behavior is very uncommon, the fact that both nodes have this behavior implies a strong network effect. Therefore, introducing exogenous factors helps our algorithm better calibrate network weights.

Model Training: During the training phase, we want to estimate the optimal values for the $\alpha_1, \ldots, \alpha_M$ and $s_1, \ldots, s_u$

We formalize this as an optimization problem by maximizing the sum of all conditional likelihoods. This is a concave optimization problem. Therefore, global optimal is guaranteed, and there exist efficient algorithms scalable to larger data sets.

Experimental Results: In experiments such as predicting mobile phone app adoption, this method yields predictions of future app adoptions with about five times the accuracy of a Bayesian estimate using population statistics.[15] We emphasize that our algorithm doesn't address the causality problem in network effects, i.e., we don't attempt to understand the different reasons why network neighbors have similar behaviors. It could be either diffusion (e.g., my neighbor tells me), or homophily (e.g., network neighbors share the same interests and personality), or a common third cause.

Trend Prediction in Social Networks (Chapter 2)

Given observations of exposure in a social network, one should be able to calculate the probability of a new behavior that appears in some individuals, and then subsequently spreads to a large number of individuals. This is what I have named idea flow, that is, the spread of a new idea through a network.

One of the main difficulties with such trend prediction stems from the fact that the first spreading phase of "soon to be global trends" is quite similar to other types of network patterns. In other words, given several observed behavior changes in a social network, it is very hard to predict which of them will result in a widespread trend and which will quickly dissolve into nothing.

To address this problem, postdoctoral student Yaniv Altshuler, together with Wei Pan and I, built a method of predicting trend propagation using the composite influence model framework described above.[16] We modeled the community, or social network, as a graph G that is comprised of U (the community's members) and W (social links among them). We use n to denote the size of the network, namely |U|. In this network, we are interested in predicting the future behavior of some observed anomalous pattern a. Notice that a can refer to a growing use of some new Web service such as Groupon or alternatively a behavior, such as associating oneself with the "99 percent" movement.

Notice that exposures to trends are transitive. Namely, an exposing user generates exposure agents that can be transmitted on the network's social links to exposed users, which can in turn transmit them onward to their friends, and so on. We therefore model a trend's exposure interactions as movements of random walking agents in a network. Every user that was exposed to a trend a generates β such agents, on average.

We assume that our network is (or can be approximated by) a scale-free network $G(n, c, \gamma)$, namely, a network of n users where the probability that user u has d neighbors follows a power law:

$$P(d) \sim c \cdot d^{-\gamma}$$

This model proves to be accurate for most of the social networks reported on in this book; interestingly, some that we think of as not having a power law distribution (like telephone call networks) can be modeled as a relatively fixed exogenously determined component together with an additive power law component. Recent studies have examined the way influence is being conveyed through social links. In the composite influence model described above, the probability that network users would install applications after being exposed to those installed by friends was tested. For some user u, this behavior was shown to be best modeled as follows:

$$P_{Local-Adopt}(a, u, t, \Delta t) = 1 - exp\{-(s_v + p_a(u))\}$$

Definitions and methods for obtaining the values of s_u and $w_{u,v}$ are the same as in the behavior adoption section above. For every member $u \in U$, $s_u \geq 0$ captures the individual susceptibility of this member, regardless of the specific behavior (or trend) in question. $p_a(u)$ denotes the network potential for the user u with respect to the trend a and is defined as the sum of network agnostic social weights of the user u with the friends exposing him with the trend a. Notice also that both properties are trend agnostic. However, while s_u is evaluated once for each user and is network agnostic, $p_a(u)$ contributes network-specific information, and can also be used by us to decide the identity of the network's members that we would target in our initial campaign. From $p_{Local-Adopt}$ we can calculate estimates of p_{Trend}, which I have named idea flow, as described in Altshuler and Pentland.[17] We have validated the accuracy and predictive power of our model on several comprehensive data sets, namely the Friends and Family data set that studied the social aspects of a small community of young families and the eToro data set, a set of financial transactions from 1.6 million users of a social trading community, among others. This same framework has also been used to model idea flow in both companies and entire cities, and to relate idea flow to productivity and GDP, as will be described in the next section.

Idea Flow in Companies and Cities (Chapters 6, 9)

By using sociometric badges, we can measure interactions within companies, and by using mobile phones, we can generate a good model of social-tie density in cities. By combining parameters from specific examples of behavior propagation (e.g., app adoption or purchasing patterns) with the topology of these networks, we can build a quantitative model of how ideas flow across these particular social networks. We can then simulate how new ideas turn into new behaviors, and thus propagate throughout the network.

To accomplish this mathematical simulation task, we have to recall that people have two ways of thinking: fast and slow (see Chapter 3 and Appendix 3 for refreshers). This also gives people two ways of learning.

For the slow mode, often a single exposure to a new idea or a new piece of information will be enough to change behavior. An example of this simple contagion model is the spreading of a new fact (that road is under construction) or a rumor ("she did what!?"). This same model is also typical of the spread of disease through a population. Infectious ideas, like infectious diseases, travel along social ties. This is simulated by a cascade of state transitions within the influence model of the social network.

We know, however, that much of our behavior is due to fast-thinking habits. Here a simple contagion model does not do a good job of capturing changes in many habitual behaviors. For the fast mode of thinking we usually need exposure to several examples in which someone else successfully used the new behavior before we are willing to try it for ourselves. In these cases, a second, complex contagion, model is a better description of the adoption of habitual, fast-thinking behaviors.

This is exactly what we saw in Chapter 3 with the adoption of new social network technologies and new mobile apps, and it is also how exposure drives changes in eating habits, political views, and more. It is simulated by a cascade of state transitions within the influence model of the social network but now with the network parameters set to match this more conservative type of idea spread.

So to connect information and idea flow along social ties to changes in behavior, we must therefore account for both fast and slow thinking. Mathematically, this means that we must examine two different influence models. In one version we will use a simple contagion assumption, in which only one exposure to an idea is enough to cause behavior change. In the second case, when multiple exposures to the same idea are required before a person adopts the new behavior, we will use a complex contagion assumption.

The two models have different values of p_{Trend}, our measure of idea flow that predicts how likely an idea is to spread throughout a community. The two models have only one real difference, however: the number of positive examples that are required within a short period of time before behavior change occurs. As a result, for ideas that are introduced into the social network repeatedly over long periods of time, the models generate quite similar patterns of spreading behavior change. The big difference between them is that with the complex model, new behaviors spread much more slowly, and behavior changes often do not reach the thinly linked boundaries of the social network. For many applications, such as modeling GDP, the speed difference between the simple and complex contagion models is not an issue, because we are comparing stable, steady state situations.

Social Pressure (Chapter 4)

Cooperation in large societies of self-interested individuals is a crucial yet extremely difficult goal to achieve.[18] Some of the most important problems in modern society, such as pollution, global warming, rising health care and insurance costs, among others, arise from an inability to achieve cooperation at a large scale.

A tragedy of the commons occurs when multiple individuals, acting rationally in their own self-interest, will ultimately deplete a common resource to the detriment of everyone.[19] The cause of the tragedy is that the negative externality from any individual's uncooperative action is experienced by the larger society, yet the benefit accrues entirely to the individual.

The scientific literature suggests that cooperation is much more easily achieved locally, among peers, than among anonymous individuals.[20] If an individual's actions affect only their peers, not only do her peers experience a negative externality as a result of the uncooperative action, but she also incurs a social cost. One way in which the peers achieve cooperation among themselves is through costly peer pressure.[21]

Traditional solutions to the problem of cooperation in large societies are quotas and taxes/subsidies. Quotas enforce limits on the production of negative externalities, while a more market-based approach is Pigouvian (individual) taxation or subsidy.[22] There are two reasons why subsidies are better than taxes: a) subsidies give positive feedback and have a better effect; and b) it is harder to institute a public policy in a free society that taxes people if they do not take a cooperative action (such as lead a healthy lifestyle).[23] Similar to the impact of Pigouvian taxes, subsidies cause individuals to internalize the externalities resulting from their actions.

In effect, these policies tax everyone in the society and redistribute it as subsidies to enforce cooperation. The budget required for the subsidies can be substantial and may include significant redistribution overhead. The outcomes resulting from such policies are not optimal for the society. There are two problems: a) the Coasian argument fails here, due to the presence of huge transaction costs, and so a simple redistribution does not achieve a Pareto-efficient outcome; and b) these policies assume that the society consists of a population of independent individuals and discounts the fact that individual decisions are influenced by interactions with their peers.[24] That is, the standard model for externalities does not account for interactions between peers in the society.

Ankur Mani, in his PhD thesis with me, also worked with visiting Masdar faculty member Iyad Rahwan to introduce interactions between peers into the tragedy of the commons problem, adding a new, joint model of externalities and peer interactions as peer pressure.[25] We proposed a new model for networked societies and provided a new set of mechanisms for policy makers to address the problem of externalities.

These mechanisms are suitable for a networked society in which externalities are global but interactions are local. Rather than the individual internalizing the externalities via Pigouvian taxation or subsidies, we localize them to one's peers in a social network, thus leveraging the power of peer pressure. When the externalities are localized, cooperation is achieved locally, and thus global cooperation is also observed. Therefore, the social mechanisms incentivize peers (via taxation or subsidy to them) to exert pressure (positive or negative) on an individual, thus causing a drop in negative externality (or increase in positive externality).

We show that under certain very general conditions, this approach can yield a socially efficient and better outcome at a lower budget than the one for the Pigouvian subsidies.

Our main insight is that by targeting the individual's peers, peer pressure can amplify the desired effect of a reward on the target individual. In contrast with the Pigouvian approach, which focuses on the individual causing the externality, our mechanism focuses on their peers in the social network. The idea is to incentivize agent A's peers to exert (positive or negative) pressure on A.

Our mechanism can be summarized by the following question: If we reward the peers of agent A, can we encourage them to exert more pressure on A to reduce the negative externality? And is this policy efficient compared to Pigouvian policies?

By targeting the individual's peers, peer pressure can amplify the desired effect on the target individual. That is, under certain conditions, the resulting reduction in negative externality can be larger, given an identical subsidy budget.

We studied a joint strategic model of externalities and peer pressure in social networks in which agents take actions that exert externalities on the whole network and also apply costly peer pressure on their peers. This model is closely related to the formulation of Calvó-Armengol and Jackson.[26]

It turns out that in the equilibrium of this game, only the peers who feel the highest externality apply pressure. Furthermore, the pressure felt by any individual in the network is the same in all the equilibria. It yields some improvement in social surplus, but may not be optimal.

With these characterization results, we then explored how, using information about the structure of the social network, optimal social surplus can be achieved using carefully designed social mechanisms. We were able to show that the social mechanism achieves the optimal outcome at a lower budget and total cost than the Pigouvian mechanism.

Social mechanisms are superior for two reasons: 1) when all the externalities are internalized as in the Pigouvian mechanism, then there is no peer pressure on the agent creating the externality, and thus requires additional subsidies; and 2) when the marginal cost of exerting peer pressure is lower than the marginal externality on the whole society times the marginal response to peer pressure, then the effect of subsidies is amplified in the social mechanism. This increases with the strength of the relationships between the peers and is inversely proportional to the cost of exerting peer pressure.

We anticipate two applications of such mechanisms: 1) public policy for reducing global externalities, such as pollution; and 2) revenue maximization for products with network externalities, such as collaborative search engines or social recommendations.

Externalities with peer pressure: In this new model, actors have the ability to exert pressure on their peers in the social network. The utility U of all actors x in network p is defined by the individual utilities u_i, the cost of externalities v_i that are exerted on i by other individuals, the cost c of exerting peer pressure on individual x_i by their neighbors, together with r_{ji}, the social network incentive, and p_{ij}, the peer pressure exerted by the agent i on their peer j. Note that if i and j are not peers in the social network, then $p_{ij} = 0$.

$$U_i(x, p) = u_i(x_i) - v_i \left(\sum_{j \neq i} x_j \right) - x_i \sum_{j \in Nbr(i)} p_{ji} - c_i \sum_{j \in Nbr(i)} p_{ji} + \sum_{j \in Nbr(i)} r_{ji}(x_j)$$

It is assumed that u is strictly concave, and v is strictly convex, and increasing. In terms of the influence model, the incentives are boundary conditions that alter the state evolution of the entire social network. What the social network incentives r_{ji} accomplish is to cause agent i to choose to modify its state transition probabilities so that it is more likely to assume states that influence its neighboring agents j to adopt the desired behavior. That is, the incentive encourages agent i to apply social pressure to agent j.

Social network mechanisms (rewarding the peers): As described earlier, the social mechanism rewards individuals for their peers' actions, in effect subsidizing the cost of the peer pressure they exert. There are many possible reward structures for creating social mechanisms. We discuss here one in which the reward is given to agent i as a result of her peer agent j's action x_j.

What might be a suitable approach to allocating such social reward? We would like to identify a reward function that has the following properties:

1. The reward must be simple. We consider reward functions with a constant marginal reward (i.e., affine reward functions).
2. A subgame perfect equilibrium of the game should exist.
3. Equilibrium action should be optimal.
4. Each peer is rewarded for an agent's reduced action.
5. Budgets for rewards should be minimized over the set of reward functions that satisfy the above conditions.

It turns out that a fairly simple reward function satisfies these conditions. Changing conditions 1–5 may give different reward functions. Note that this reward has a component that depends upon the consumer and a component that depends upon the neighbor.

Summary

This appendix has sketched the mathematics used in the various examples in this book, but of course those with a serious interest should turn to the original material. Further, it is not the intent of this appendix to present *the* mathematics of social physics, even though this particular method for data-driven modeling of social network phenomena has proven quite accurate and robust. I am sure that better formulations will be developed over time.

The main point is that the propagation of human action habits by means of social learning can be accurately modeled from easily observable behavior using heterogeneous, dynamic, stochastic networks. This capability is transformative for increasing our understanding of the dynamics of human society, and hence our ability to plan for our future.

Notes

Chapter 1: From Ideas to Action

1. A. Smith 2009.
2. In more technical language, the time has come to consider dynamics and not just equilibria, and exchange networks rather than just pooled markets. In addition, we must consider social influence together with rationality and view utility as a vector (e.g., fitness, curiosity, status, etc.) rather than as a scalar.
3. Zipf 1949.
4. Zipf 1946.
5. Snijders 2001; Krackhardt and Hanson 1993; Macy and Willer 2002; Burt 1992; Uzzi 1997; White 2002.
6. Kleinberg 2013; Barabási 2002; Monge and Contractor 2003; Gonzalez et al. 2008; Onnela et al. 2007, 2011.
7. Centola 2010; Lazer and Friedman 2007; Aral et al. 2009; Eagle et al. 2010; Pentland 2008.
8. Marr 1982.
9. Pentland 2012c, 2013a.
10. Lazer et al. 2009.
11. Barker 1968; Dawber 1980.
12. There were dozens of standard psychological, sociological, and health-related surveys administered regularly in these living labs, usually on the Web. In addition, there were also shorter, more frequent questionnaires administered on the smartphones.
13. Aharony et al. 2011.
14. Madan et al. 2012.
15. Eagle and Pentland 2006.
16. Pentland 2012b.

17. Participants had the protections of informed consent, the ability to withdraw at any time, and guaranteed secure handling of all personal data, and they were paid to participate.
18. Pentland 2009.
19. World Economic Forum 2011. "Personal Data: The Emergence of a New Asset Class." See http://www3.weforum.org/docs/WEF_ITTC_PersonalDataNewAsset_Report_2011.pdf.
20. Along with small experiment sizes, almost all social science is based on people from Western, educated, international, rich, democratic societies. In other words, social science is just for the WEIRD (Henrich et al. 2010).
21. Kahneman 2011.

Chapter 2: Exploration

1. Beahm, George, ed. *I, Steve: Steve Jobs in His Own Words* (Chicago: Agate B2), 2011.
2. Papert and Harel 1991.
3. Buchanan 2007.
4. Conradt and Roper 2005.
5. Surowiecki 2004.
6. Dall et al. 2005.
7. Lorenz et al. 2011.
8. Dall et al. 2005; Danchin et al. 2004.
9 King et al. 2012.
10. Hong and Page 2004; Krause et al. 2011.
11. Altshuler et al. 2012; Pan, Altshuler, and Pentland 2012. eToro (http://www.etoro.com) is an online discounted retail broker for foreign exchange and commodity trading that has easy-to-use buying, short selling, and leveraging mechanisms. eToro makes financial trading both accessible and fun, as it allows any user to take long and short positions with a minimal bid of a few dollars. It's rather similar to playing the lottery, but users compete with the real world rather than a lottery computer. Despite having three million customers at the time of our study, it is important to remember that eToro remains a small player in the foreign exchange market—the traders in eToro do *not* move the market.
12. Using sophisticated mathematical analyses, we are able to calculate a measure of the rate of idea flow. The rate of idea flow is the likelihood that a certain proportion of users (expressed as a probability distribution) will adopt a new strategy that is introduced into the social network. This crucial measure takes into account the social network structure and the strength of the social influence between each person as well as the individual susceptibility to new ideas. For those who are interested in the

math, more details about how idea flow is calculated can be found in the Math appendix.

13. Subtraction of the individual trading return on investment (ROI) makes the vertical axis market neutral, since individual traders have market-neutral performance.

14. The vertical variation in ROI for a given value of rate of idea flow is due to the fact that there were different proportions of asset classes on different days. Each asset class has its own slightly different sweet spot in terms of the rate of idea flow, and when this variation is taken into account, the variation in ROI is dramatically reduced.

15. Yamamoto et al. 2013; Sueur et al. 2012.

16. Farrell 2011.

17. Lazer and Friedman 2007.

18. Glinton et al. 2010; Anghel et al. 2004.

19. One where the probability that a user had d followers is $Prob(d) \sim d^{-\gamma}$

20. Shmueli et al. 2013. That is, the number of connections that change have a wide range of change sizes.

21. In Chapter 4 we will see another critical difference between infection and behavior change. While attentive, conscious beliefs ("the store opens at 8:00 A.M.") can be spread with just one comment, it turns out that people usually require several role model examples within a short period of time before adopting a new behavior to replace habitual, largely unconscious behaviors (for instance, using cash rather than a credit card). The first type of behavior change is known as simple contagion, and the second as complex contagion. These two types of behavior change spread through a network very similarly, but the spread of complex contagion is much slower and typically requires a densely connected local network, so that when an idea enters a person's immediate social network, they will have many exposures to the idea within a short period of time. See Watts and Dodds 2007; Centola 2010; Centola and Macy 2007.

22. Kelly 1999.

23. Choudhury and Pentland 2004.

24. In technical terms, individuals who were more influential in conversational turn taking also had higher betweenness centrality in the social network. This was an extremely strong relationship, with an r^2 of 0.9.

25. Pan, Altshuler, and Pentland 2012; Saavedraa et al. 2011.

26. *Financial Times*, April 18, 2013.

27. It is important to diversify by considering more than one strategy at a time because, as the environment changes, the old strategies stop working and new strategies take the lead. Therefore, it is not the strategies that *have been* most successful that you want to follow; it is really the strategies that *will be* most successful that you have to find. Since predicting the future is hard, diversification of social learning is important.

Chapter 3: Idea Flow

1. Bandura 1977.
2. Meltzoff 1988.
3. Perhaps ape "culture" is like the stagnant cultures of isolated villages and tribes, in which idea sharing is only with a closed group and so community behavior remains rigid and uncreative.
4. In the Social Evolution experiment some data required significant postprocessing. For instance, if your phone sensed me but mine did not sense you, we would mark both phones as proximate to each other. Similarly, if two phones sensed a WiFi hot spot, they would be marked as being in the same area. The Friends and Family study had better sensing and did not require this postprocessing. For more detail see http://realitycommons.media.mit.edu.
5. Christakis and Fowler 2007.
6. Madan et al. 2012.
7. In this chapter I report on health habits, political views, app adoption, and music downloads, all of which show similar mechanisms and effect magnitudes. In the next chapter I report on manipulations that change health habits and consumer, voting, and office behavior (use of digital social networks).
8. Social influence is an active and contentious area of research. (Aral et al. 2009). The health, politics, and app adoption studies in this chapter are stronger than most studies because: 1) these effects seem primarily to be social learning, not social pressure. The behavior of those with strong ties (e.g., friends) had no significant effect, while exposure to acquaintances with very weak ties has a strong effect; 2) we have a sequence of measurements rather than a single snapshot so that we can determine if the timing is appropriate for cause and effect; and 3) we have quantitative, continuous measurements of exposure, not just binary indications of social ties. Finally, our real-world results are quite similar to those found in online experiments such as those of Damon Centola (Centola 2010), where the circumstances can be precisely controlled.
9. Madan et al. 2011.
10. This was a temporary effect, however, because after the political debates things went back to normal.
11. Aharony et al. 2011.
12. Pan et al. 2011a.
13. Krumme et al. 2012; Tran et al. 2011.
14. Salganik et al. 2006.
15. Rendell et al. 2010.
16. Lazer and Friedman 2007; Glinton et al. 2010; Anghel et al. 2004; Yamamoto et al. 2013; Sueur et al. 2012; Farrell 2011.

17. Simon 1978; Kahneman 2002.
18. Kahneman 2011.
19. Hassin et al. 2005.
20. Rand et al. 2009; Fudenberg et al. 2012.
21. Haidt 2010.
22. Brennan and Lo 2011.
23. Hassin et al. 2005.

Chapter 4: Engagement

1. Stewart and Harcourt 1994.
2. Boinski and Campbell 1995.
3. Conradt and Roper 2005; Couzin et al. 2005; Couzin 2007.
4. Kelly 1999.
5. Cohen et al. 2010.
6. Calvó-Armengol and Jackson 2010.
7. Kandel and Lazear 1992.
8. Breza 2012.
9. Nowak 2006.
10. Rand et al. 2009; Fehr and Gachter 2002.
11. Pink 2009; Gneezy et al. 2011.
12. Mani, Rahwan, and Pentland 2013.
13. That is, four times the behavior change per dollar of incentive.
14. The marginal cost for a unit of improvement in each of these three conditions was even more impressive:
    Individual (Pigouvian) incentives: $83
    Peer view: $39.5
    Peer reward: $12
    Similarly, the mean percent improvement in activity is also impressive:
    Individual incentives: 3.2 percent
    Peer view: 5.5 percent
    Peer reward: 10.4 percent
15. Adjodah and Pentland 2013.
16. E.g., conversations and phone calls, but not indirect interactions such as overhearing or observing the other person.
17. The correlation of the amount of behavior change with the number of phone calls was $r^2 > 0.8$; for all channels of communication, $r^2 > 0.9$.
18. We asked trust questions of every pair of people in the community such as: Would they trust the other person to babysit their child? Would they loan them money? Would they loan them their car? At the end we counted the number of yes answers for each pair of people and called it their trust score. When postdoctoral students Erez Shmueli and Vivek Singh and I compared their trust score to the number of times they had

direct interactions with one another we found that the total amount of direct interaction gave a surprisingly accurate prediction of the trust score. Again, r^2 > 0.8 for phone calls, r^2 > 0.9 for all channels of communication.

19. Mani et al. 2012.
20. Mani, Loock, Rahwan, and Pentland 2013.
21. De Montjoye et al. 2013.
22. Smith 2009.
23. Lim et al. 2007.
24. Nowak 2006; Rand et al. 2009; Fehr and Gachter 2002.
25. Buchanan 2007.
26. Stewart and Harcourt 1994; Boinski and Campbell 1995.
27. Zimbardo 2007; Milgram 1974b.
28. Pentland 2008; Olguín et al. 2009; Pentland 2012b.
29. Dong and Pentland 2007; Pan, Dong, Cebrian, Kim, Fowler, and Pentland 2012.
30. Castellano et al. 2009; Gomez-Rodriguez et al. 2010.
31. Dong et al. 2007; Pan, Dong, Cebrian, Kim, Fowler, and Pentland 2012.

Chapter 5: Collective Intelligence

1. Woolley et al. 2010.
2. Pentland 2011.
3. Dong et al. 2009; Dong et al. 2012; Pentland 2008.
4. Pentland 2010a; Cebrian et al. 2010.
5. Olguín et al. 2009; see also www.sociometricsolutions.com.
6. Pentland 2012b. This paper won both the McKinsey Award from the *Harvard Business Review* and the Academy of Management's Practitioner Award.
7. Wu et al. 2008.
8. Couzin 2009.
9. Ancona et al. 2002.
10. Olguín et al. 2009.
11. Eagle and Pentland 2006.
12. Dong and Pentland 2007.
13. Amabile et al. 1996.
14. Tripathi 2011; Tripathi and Burleson 2012.
15. Hassin et al. 2005.
16. This is also called network constraint.
17. Pentland 2012b.

Chapter 6: Shaping Organizations

1. Pentland 2012b.
2. As I hope is clear by now, each interaction or exposure is a learning opportunity, and our experimental results indicate that effective idea flow from one person to another, e.g., the probability of adopting a new behavior is a smooth increasing function of the number of interactions and exposures. It is worth noting that this is consistent with the findings of sociology pioneers such as Ron Burt, who focused mainly on the network topology and the frequency of communication. If you are a cognitive scientist, you will probably be uncomfortable with the idea that the relationship between exposure and idea adoption has such a simple form. Those are the data, however: Statistically there is a fairly uniform average adoption rate that we can easily calculate. Note, though, that different sorts of ideas have different propagation properties, different channels of communication have different influence characteristics, and individuals have different susceptibilities. If you are a computer scientist, you may also worry that exposure (proximity) is being conflated with communication; however, I have been careful to specify which is which. In addition, see Wyatt et al. 2011, in which they examine the relationship between proximity and the likelihood of conversation. While the paper makes it clear that these are separate phenomena, it is also clear that if we observe all the members of a population over periods of a week or more, then frequency of conversation and frequency of proximity are highly correlated. See Chapter 4, the Fast, Slow, and Free Will appendix, and the Math appendix for more details.
3. Burt 2004.
4. Kim et al. 2008; Kim 2011.
5. In the context of a meeting, engagement means that everyone is both contributing ideas and responding to all the other people's ideas. In other words, it should not be the same people who always respond to a particular speaker.
6. In these experiments, trust is measured via a classic public goods game.
7. Kim 2011.
8. See "Sensible Organization: Inspired by Social Sensor Technologies" at http://hd.media.mit.edu/tech-reports/TR-602.pdf.
9. Wellman 2001.
10. Pentland 2012b; see also www.sociometricsolutions.com.
11. Chen et al. 2003; Chen et al. 2004.
12. Prelec 2004.
13. More technically, keep track of the conditional probabilities between people. This can be accomplished by the use of the influence model described in the Math appendix.

14. We have demonstrated that these are causal relationships; see Kim 2011.
15. Pentland 2010b.
16. Choudhury and Pentland 2003, 2004.

Chapter 7: Organizational Change

1. Pickard et al. 2011.
2. Rutherford et al. 2013.
3. See http://archive.darpa.mil/networkchallenge.
4. Nagar 2012.
5. Olguín et al. 2009.
6. Waber 2013.
7. Wellman 2001.
8. Putnam 1995.
9. Pentland 2008.
10. Buchanan 2009.
11. Lepri et al. 2009; Dong et al. 2007.
12. Curhan and Pentland 2007.
13. Choudhury and Pentland 2004.
14. Barsade 2002.
15. Iacoboni and Mazziotta 2007.

Chapter 8: Sensing Cities

1. Pentland 2012a.
2. See www.sensenetworks.com.
3. Eagle and Pentland 2006.
4. Dong and Pentland 2009.
5. Berlingerio et al. 2013.
6. Smith, Mashadi, and Capra 2013.
7. Schneider 2010.
8. Madan et al. 2010; Madan et al. 2012; Dong et al. 2012.
9. See www.ginger.io.
10. Dong et al. 2012; Pentland et al. 2009.
11. Dong et al. 2012.
12. Lima et al. 2013; Pentland et al. 2009.
13. Mani, Loock, Rahwan, and Pentland 2013.
14. Pentland 2012a.
15. Lima et al. 2013; Smith, Mashadi, and Capra 2013; Berlingerio et al. 2013; Pentland et al. 2009; Pentland 2012a.

Chapter 9: City Science

1. Crane and Kinzig 2005.
2. Glaeser et al. 2000.

3. Smith 1937.
4. Milgram 1974a; Becker et al. 1999; Krugman 1993; Fujita et al. 1999; Bettencourt et al. 2007; Bettencourt and West 2010.
5. Audretsch and Feldman 1996; Jaffe et al. 1993; Anselin et al. 1997.
6. Arbesman et al. 2009; Leskovec et al. 2009; Expert et al. 2011; Onnela et al. 2011; Mucha et al. 2010.
7. Pan et al. 2013.
8. Krugman 1993.
9. Wirth 1938; Hägerstrand 1952, 1957; Florida 2002, 2005, 2007.
10. Liben-Nowell et al. 2005.
11. Along with this smoothly decreasing function of relationship frequency that seems to be a natural result of face-to-face interactions, there is also a baseline of about two fifths of all relationships that is independent of distance and may well originate from online introductions. This suggests that digital communications are changing the relationship between social ties and city productivity/creativity. It is important to remember, however, that face-to-face social ties are far more important than digital ties when it comes to changing habits, which means that we may have increasing exploration but only slowly increasing behavior change.
12. Nguyen and Szymanski 2012.
13. $P_j = 1/\text{rank}(j)$, essentially the chance you have a social tie is inversely proportional to the number of intervening other people.
14. U.S. Centers for Disease Control; see http://www.cdc.gov/hiv/topics/surveillance/index.htm.
15. Calabrese et al. 2011.
16. Krumme 2012.
17. Krumme et al. 2013.
18. It is Zipf's law, named after the fellow who discovered this law in other social phenomena.
19. Pan et al. 2011b.
20. Frijters et al. 2004; Paridon et al. 2006; Clydesdale 1997; Pong and Ju 2000.
21. I take average commuting distance to be half the best-fit maximum radius of interaction calculated from GDP.
22. Smith, Mashadi, and Capra 2013; Smith, Quercia, and Capra 2013.
23. Increased crime, similar to increased productivity, seems to be a product of innovation.
24. Jacobs 1961.
25. I assume that there are six major peer groups: men's groups and women's groups, each with subgroups of young, parents, and seniors. Each peer group is a Dunbar number (150) squared, which will be the maximum number of friends of friends.

26. To be critical, what we are talking about is creating places where there is great social support but where change happens slowly. I would argue that this tends to protect children and families against fast and destructive change, which is a real and increasing danger in the emerging hyperconnected world. Others will, of course, disagree and prefer a greater rate of social change.

27. Burt 1992; Granovetter 1973, 2005; Eagle et al. 2010; Wu et al. 2008; Allen 2003; Reagans and Zuckerman 2001.

28. Eagle and Pentland 2009; Wu et al. 2008; Pentland 2008.

29. Kim et al. 2011.

30. Singh et al. in preparation.

Chapter 10: Data-Driven Societies

1. World Economic Forum 2011. "Personal Data: The Emergence of a New Asset Class." See http://www3.weforum.org/docs/WEF_ITTC_Per sonalDataNewAsset_Report_2011.pdf.

2. Pentland 2009.

3. Ostrom 1990.

4. De Soto and Cheneval 2006.

5. Pentland 2009.

6. World Economic Forum 2011, "Personal Data: The Emergence of a New Asset Class." See http://weforum.org/docs/WEF_ITTC_Personal DataNewAsset_Report_2011.pdf.

7. See www.idcubed.org.

8. De Montjoye et al. 2012.

9. Smith, Mashadi, and Capra 2013; Bucicovschi et al. 2013.

10. De Montjoye et al. 2012.

Chapter 11: Design for Harmony

1. Smith 2009.

2. Nowak 2006; Rand et al. 2009; Ostrom 1990; Putnam 1995.

3. Weber 1946.

4. Marx 1867.

5. Acemoglu et al. 2012.

6. The international economy has a similarly constrained network structure; see Hidalgo et al. 2007.

7. Salamone 1997; Lee 1988; Gray 2009; Thomas 2006.

8. Mani et al. 2010.

9. Localized social efficiency arises because each participant in the network finds the best exchange available in the part of the network they are connected to (e.g., they search for a Pareto optimal exchange). By construc-

tion, this yields social optimality subject to constraints imposed by the topology of the network. For proof of convergence, see Mani et al. 2010.

10. See also Bouchaud and Mezard 2000.

11. See also Grund et al. 2013, and Helbing et al. 2011 for similar results.

12. Fair exchange networks are also stable with respect to coalitions of people. Such coalitions arise when a peer group (for instance, bankers) develops a social norm about how to deal with other people (for instance, lawyers), and their shared habits allow them to act in concert. A network society can be stable and fair even with peer groups acting in concert with each other, because such coalitions can be balanced by the shared habits of the peer groups they trade with. Mathematically, the exchange network comes to include "super nodes" that consist of peer groups rather than individuals, but this does not destroy the fairness and trust properties of the society.

13. Lim et al. 2007.

14. Dunbar 1992.

15. That is, when most people have satisfied (maxed out) their utility function.

16. Information is the stuff from which ideas can potentially be created, and in addition, it is an important source of our beliefs.

17. See http://www.swift.com.

18. Rand et al. 2009; Sigmund et al. 2010.

19. Smith, Mashadi, and Capra 2013.

20. Eagle et al. 2010.

21. Bucicovschi et al. 2013.

22. Berlingerio et al. 2013.

23. Lima et al. 2013.

Appendix 1: Reality Mining

1. Lazer et al. 2009.

Appendix 2: OpenPDS

1. World Economic Forum 2011. "Personal Data: The Emergence of a New Asset Class." See http://www3.weforum.org/docs/WEF_ITTC_Per sonalDataNewAsset_Report_2011.pdf.

2. See http://idcubed.org.

3. De Montjoye et al. 2012.

4. Pentland 2009.

5. National Strategy for Trusted Identities in Cyberspace. "*National Strategy for Trusted Identities in Cyberspace*" initiative. See: http://www.nist .gov/nstic.

6. International Strategy for Cyberspace. See: http://www.whitehouse.gov/sites/default/files/rss_viewer/international_strategy_for_cyberspace.pdf.
7. "Commission Proposes a Comprehensive Reform of Data Protection Rules to Increase Users' Control of Their Data and to Cut Costs for Businesses." See: http://europa.eu/rapid/press-release_IP-12-46_en.htm.
8. World Economic Forum 2011, "Personal Data: The Emergence of a New Asset Class." See http://www3.weforum.org/docs/WEF_ITTC_PersonalDataNewAsset_Report_2011.pdf.
9. "Has Big Data Made Anonymity Impossible?" See: http://www.technologyreview.com/news/514351/has-big-data-made-anonymity-impossible.
10. Sweeney 2002.
11. Schwartz 2003; Butler 2007; "Your Apps Are Watching You." See: http://online.wsj.com/article/SB10001424052748704694004576020083703574602.html.
12. Blumberg and Eckersley 2009.
13. See: http://www.darpa.mil/Our_Work/I2O/Programs/Detection_and_Computational_Analysis_of_Psychological_Signals_(DCAPS).aspx.

Appendix 3: Fast, Slow, and Free Will

1. Kahneman 2002; Simon 1978.
2. Centola 2010; Centola and Macy 2007.
3. Slow thinking is not nearly as good as we would like to believe. For instance, Tetlock 2005 shows that the best world experts are little better than chance at making predictions, even in their own field of expertise.
4. Dijksterhuis 2004.
5. Hassin et al. 2005.
6. Kahneman 2011.
7. Lévi-Strauss 1955; Marx 1867; Smith 1937; Sartre 1943; Arrow 1987.
8. Kahneman 2002, 2011; Hassin et al. 2005; Pentland 2008; Simon 1978; Bandura 1977.

Appendix 4: Math

1. Dong and Pentland 2007.
2. Pan, Dong, Cebrian, Kim, Fowler, and Pentland 2012.
3. Granovetter and Soong 1983.
4. Aral et al. 2009.
5. Gomez-Rodriguez et al. 2010; Myers and Leskovec 2010.
6. Dong and Pentland 2007.
7. Lepri et al. 2009.
8. Dong and Pentland 2009.
9. Dong et al. 2012.
10. Pan, Dong, Cebrian, Kim, Fowler, and Pentland 2012.

11. Ibid.
12. Pan et al. 2011a.
13. Christakis and Fowler 2007.
14. Centola 2010.
15. Pan et al. 2011a.
16. Altshuler et al. 2012.
17. Ibid.
18. Dietz et al. 2003.
19. Hardin 1968.
20. Baumol 1972.
21. Calvó-Armengol and Jackson 2010.
22. Baumol 1972; Slemrod 1990.
23. Nowak 2006.
24. Coase 1960.
25. Mani, Rahwan, and Pentland 2013.
26. Calvó-Armengol and Jackson 2010.

References

Acemoglu, D., V. Carvalho, A. Ozdaglar, and A. Tahbaz-Salehi. 2012. "The Network Origins of Aggregate Fluctuations." *Econometrica* 80 (5): 1977–2016.

Adjodah, D., and A. Pentland. 2013. "Understanding Social Influence Using Network Analysis and Machine Learning." *NetSci Conference*, Copenhagen, Denmark, June 5–6.

Aharony, N., W. Pan, I. Cory, I. Khayal, and A. Pentland. 2011. "Social fMRI: Investigating and Shaping Social Mechanisms in the Real World." *Pervasive and Mobile Computing* 7, no. 6 (December): 643–59.

Allen, T. 2003. *Managing the Flow of Technology: Technology Transfer and the Dissemination of Technological Information Within the R&D Organization*. Cambridge, MA: MIT Press.

Altshuler, Y., W. Pan, and A. Pentland. 2012. "Trends Prediction Using Social Diffusion Models." In *Social Computing, Behavioral-Cultural Modeling and Prediction*. Lecture Notes in Computer Science series. 7227 Berlin, Heidelberg: Springer. 97–104.

Amabile, T. M., R. Conti, H. Coon, J. Lazenby, and M. Herron. 1996. "Assessing the Work Environment for Creativity." *Academy of Management Journal* 39 (5): 1154–84.

Ancona, D., H. Bresman, and K. Kaeufer. 2002. "The Comparative Advantage of X-teams." *MIT Sloan Management Review* 43, no. 3 (Spring): 33–40.

Anghel, M., Z. Toroczkai, K. Bassler, and G. Korniss. 2004. "Competition in Social Networks: Emergence of a Scale-Free Leadership Structure and Collective Efficiency." *Physical Review Letters* 92 (5): 058701.

Anselin, L., A. Varga, and Z. Acs. 1997. "Local Geographic Spillovers Between University Research and High Technology Innovations." *Journal of Urban Economics* 42 (3): 422–48.

Aral, S., L. Muchnik, and A. Sundararajan. 2009. "Distinguishing

Influence-Based Contagion from Homophily-Driven Diffusion in Dynamic Networks." *Proceedings of the National Academy of Sciences* 106 (51): 21544–49.

Arbesman, S., J. Kleinberg, and S. Strogatz. 2009. "Superlinear Scaling for Innovation in Cities." *Physical Review E* 79 (1): 16115.

Arrow, K. J. 1987. "Economic Theory and the Hypothesis of Rationality." In *The New Palgrave: Utility and Probability*, ed. J. Eatwell, M. Milgate, and P. Newman. New York: W.W. Norton (1990), 25–37.

Audretsch, D., and M. Feldman. 1996. "R&D Spillovers and the Geography of Innovation and Production." *The American Economic Review* 86 (3): 630–40.

Bandura, A. 1977. *Social Learning Theory*. Englewood Cliffs, NJ: Prentice-Hall.

Barabási, A.-L. 2002. *Linked: The New Science of Networks*. Cambridge, MA: Perseus Publishing.

Barker, R. 1968. *Ecological Psychology: Concepts and Methods for Studying the Environment of Human Behavior*. Palo Alto, CA: Stanford University Press.

Barsade, S. 2002. "The Ripple Effect: Emotional Contagion and Its Influence on Group Behavior." *Administrative Science Quarterly* 47 (4): 644–75.

Baumol, W. J. 1972. "On Taxation and the Control of Externalities." *The American Economic Review* 62 (3): 307–22.

Beahm, George, ed. 2011. *I Steve: Steve Jobs in His Own Words*. Chicago, Agate B2.

Becker, G., E. Glaeser, and K. Murphy. 1999. "Population and Economic Growth." *The American Economic Review* 89 (2): 145–49.

Berlingerio, M., F. Calabrese, G. Di Lorenzo, R. Nair, F. Pinelli, and M. L. Sbodio. 2013. "AllAboard: A System for Exploring Urban Mobility and Optimizing Public Transport Using Cellphone Data." See www.d4d .orange.com/home.

Bettencourt, L., J. Lobo, D. Helbing, C. Kuhnert, and G. West. 2007. "Growth, Innovation, Scaling, and the Pace of Life in Cities." *Proceedings of the National Academy of Sciences* 104 (17): 7301–6.

Bettencourt, L., and G. West. 2010. "A Unified Theory of Urban Living." *Nature* 467 (October 21): 912–13.

Blumberg, A., and P. Eckersley. 2009. "On Locational Privacy and How to Avoid Losing It Forever." San Francisco: Electronic Frontier Foundation. See https://www.eff.org/wp/locational-privacy.

Boinski, S., and A. F. Campbell. 1995. "Use of Trill Vocalizations to Coordinate Troop Movement Among White-Faced Capuchins: A Second Field Test." *Behaviour* 132 (11–12): 875–901.

Bouchaud, J. P., and M. Mezard. 2000. "Wealth Condensation in a Simple Model of Economy." *Physica A: Statistical Mechanics and Its Applications* 282 (3): 536–45.

Brennan, T., and A. Lo. 2011. "The Origin of Behavior." *Quarterly Journal of Finance* 1 (1): 55–108. See http://ssrn.com/abstract=1506264.

Breza, E. 2012. Essays on Strategic Social Interactions: Evidence from Microfinance and Laboratory Experiments in the Field. PhD thesis. Economics Department, MIT.

Buchanan, M. 2007. *The Social Atom: Why the Rich Get Richer, Cheaters Get Caught, and Your Neighbor Usually Looks Like You.* New York: Bloomsbury.

———. 2009. "Secret Signals". *Nature* 457 (January 29): 528–30.

Bucicovschi, O., R. W. Douglass, D. A. Meyer, M. Ram, D. Rideout, and D. Song. 2013. "Analyzing Social Divisions Using Cell Phone Data." See http://www.d4d.orange.com/home.

Burt, R. 1992. *Structural Holes: The Social Structure of Competition.* Cambridge, MA: Harvard University Press.

———. 2004. "Structural Holes and Good Ideas." *American Journal of Sociology,* 110 (2): 349–99.

Butler, D. 2007. "Data Sharing Threatens Privacy." *Nature* 449 (October 11): 644–45.

Calabrese, F., D. Dahlem, A. Gerber, D. Paul, X. Chen, J. Rowland, C. Rath, and C. Ratti. 2011. "The Connected States Of America: Quantifying Social Radii of Influence." In *Privacy, Security, Risk and Trust (PASSAT), 2011 IEEE Third International Conference* and *2011 IEEE Third International Conference on Social Computing (SocialCom)*: 223–30.

Calvó-Armengol, A., and M. Jackson. 2010. "Peer Pressure." *Journal of the European Economic Association* 8 (1): 62–89.

Castellano, C., S. Fortunato, and V. Loreto. 2009. "Statistical Physics of Social Dynamics." *Reviews of Modern Physics* 81 (2): 591–646.

Cebrian, M., M. Lahiri, N. Oliver, and A. Pentland. 2010. "Measuring the Collective Potential of Populations from Dynamic Social Interaction Data." *Journal of Selected Topics in Signal Processing* 4 (4): 677–86.

Centola, D. 2010. "The Spread of Behavior in an Online Social Network Experiment." *Science* 329, no. 5996 (September 3): 1194–97.

Centola, D., and M. Macy. 2007. "Complex Contagions and the Weakness of Long Ties." *The American Journal of Sociology* 113 (3): 702–34.

Chen, K. Y., L. Fine, and B. Huberman. 2003. "Predicting the Future." *Information Systems Frontiers* 5 (1): 47-61.

———. 2004. "Eliminating Public Knowledge Biases in Information-Aggregation Mechanisms." *Management Science* 50 (7): 983–94.

Choudhury, T., and A. Pentland. 2003. "Sensing and Modeling Human Networks Using the Sociometer." In *Proceedings of the 7th IEEE International Symposium on Wearable Computers*: 215–22.

———. 2004. "Characterizing Social Networks Using the Sociometer." In *Proceedings of the North American Association of Computational Social and Organizational Science*, Pittsburgh, Pennsylvania, June 10–12. See http://www.cs.dartmouth.edu/~tanzeem/pubs/Choudhury_CASOS .pdf.

Christakis, N., and J. Fowler. 2007. "The Spread of Obesity in a Large Social Network over 32 Years." *New England Journal of Medicine* 357 (July 26): 370–79.

Clydesdale, T. 1997. "Family Behaviors Among Early US Baby Boomers: Exploring the Effects of Religion and Income Change, 1965–1982." *Social Forces* 76 (2): 605–35.

Coase, R. 1960. "The Problem of Social Cost." *Journal of Law and Economics* 3: 1–44.

Cohen, E. E., R. Ejsmond-Frey, N. Knight, and R. Dunbar. 2010. "Rowers' High: Behavioural Synchrony Is Correlated with Elevated Pain Thresholds." *Biology Letters* 6, no. 2 (February 23): 106–8; doi: 10.1098/rsbl.2009.0670. Epub 2009 Sep 15.

Conradt, L., and T. Roper. 2005. "Consensus Decision Making in Animals." *Trends in Ecology and Evolution* 20 (8): 449–56.

Couzin, I. 2007. "Collective Minds." *Nature* 445 (February 15): 715.

———. 2009. "Collective Cognition in Animal Groups." *Trends in Cognitive Sciences* 13 (1): 36–43.

Couzin, I., J. Krause, N. Franks, and S. Levin. 2005. "Effective Leadership and Decision-Making in Animal Groups on the Move." *Nature* 433 (February 3): 513–16.

Crane, P., and A. Kinzig. 2005. "Nature in the Metropolis." *Science* 308, no. 5726 (May 27): 1225.

Curhan, J., and A. Pentland. 2007. "Thin Slices of Negotiation: Predicting Outcomes from Conversational Dynamics Within the First Five Minutes." *Journal of Applied Psychology* 92 (3): 802–11.

Dall, S. R. X., L. A. Giraldeau, O. Olsson, J. M. McNamara, and D. W. Stephens. 2005. "Information and Its Use by Animals in Evolutionary Ecology." *Trends in Ecology and Evolution* 20 (4): 187–93; doi:10.1016/j .tree.2005.01.010.

Danchin, E., L. A. Giraldeau, T. J. Valone, and R. H. Wagner. 2004. "Public Information: From Nosy Neighbors to Cultural Evolution." *Science* 305, no. 5683 (July 23): 487–91; doi:10.1126/science.1098254.

Dawber, T. 1980. *The Framingham Study: The Epidemiology of Atherosclerotic Disease*. Cambridge, MA: Harvard University Press.

De Montjoye, Y., C. Finn, and A. Pentland. 2013. "Building Thriving

Networks: Synchronization in Human-Driven Systems." *ChASM: 2013 Computational Approaches to Social Modeling*, Barcelona, Spain (June 5–7).

De Montjoye, Y., S. Wang, and A. Pentland. 2012. "On the Trusted Use of Large-Scale Personal Data." *IEEE Data Engineering* 35 (4): 5–8.

De Soto, H., and F. Cheneval. 2006. *Swiss Human Rights Book, Volume 1: Realizing Property Rights*. Switzerland: Rüffer&Rub.

Dietz, T., E. Ostrom, and P. Stern. 2003. "The Struggle to Govern the Commons." *Science* 302, no. 5652 (December 12): 1907–12.

Dijksterhuis, A. 2004. "Think Different: The Merits of Unconscious Thought in Preference, Development and Decision Making." *Journal of Personality and Social Psychology* 87 (5): 586–98.

Dong, W., K. Heller, and A. Pentland. 2012. "Modeling Infection with Multi-Agent Dynamics." In *Social Computing, Behavioral-Cultural Modeling and Prediction*. Lecture Notes in Computer Science series. 7227. Berlin, Heidelberg: Springer. 172–79.

Dong, W., T. Kim, and A. Pentland. 2009. "A Quantitative Analysis of the Collective Creativity in Playing 20-Questions Games." In *Proceedings of the Seventh ACM Conference on Creativity and Cognition* (October 27–30): 365–66.

Dong, W., B. Lepri, A. Cappelletti, A. Pentland, F. Pianesi, and M. Zancanaro. 2007. "Using the Influence Model to Recognize Functional Roles in Meetings." In *Proceedings of the Ninth International Conference on Multimodal Interfaces* (November 12–15): 271–78.

Dong, W., and A. Pentland. 2007. "Modeling Influence Between Experts." In *Artificial Intelligence for Human Computing*. Lecture notes in Computer Science. 4451. Berlin: Springer-Verlag. 170–89. See http://link.springer.com/chapter/10.1007/978-3-540-72348-6_9#page-1.

———. 2009. "A Network Analysis of Road Traffic with Vehicle Tracking Data." In *Proceedings: AAAI Spring Symposium: Human Behavior Modeling*. 7–12.

Dunbar, R. 1992. "Neocortex Size As a Constraint on Group Size in Primates." *Journal of Human Evolution* 20 (6): 469–93.

Eagle, N., M. Macy, and R. Claxton. 2010. "Network Diversity and Economic Development." *Science* 328, no. 5981 (May 21): 1029–31. See http://www.sciencemag.org/content/328/5981/1029.full.pdf.

Eagle, N., and A. Pentland. 2006. "Reality Mining: Sensing Complex Social Systems." *Personal and Ubiquitous Computing* 10 (4): 255–68.

———. 2009. "Eigenbehaviors: Identifying Structure in Routine." *Behavioral Ecology and Sociobiology* 63 (7): 1057–66.

Expert, P., T. Evans, V. Blondel, and R. Lambiotte. 2011. "Uncovering

Space-Independent Communities in Spatial Networks." *Proceedings of the National Academy of Sciences* 108 (19): 7663–68.

Farrell, S. 2011. "Social Influence Benefits the Wisdom of Individuals in the Crowd." *Proceedings of the National Academy of Sciences* 108 (36): E625.

Fehr, E. and S. Gachter. 2002. "Altruistic Punishment in Humans." *Nature* 415 (January 10): 137–40.

Florida, R. 2002. *The Rise of the Creative Class and How It's Transforming Work, Leisure, Community, and Everyday Life.* New York: Basic Books.

———. 2005. *Cities and the Creative Class.* New York: Routledge.

———. 2007. *The Flight of The Creative Class: The New Global Competition for Talent.* New York: HarperCollins.

Frijters, P., J. Haisken-DeNew, and M. Shields. 2004. "Money Does Matter! Evidence from Increasing Real Income and Life Satisfaction in East Germany Following Reunification." *American Economic Review* 94 (3): 730–40.

Fudenberg D., D. G. Rand, and A. Dreber. 2012. "Slow to Anger and Fast to Forgive: Cooperation in an Uncertain World." *American Economic Review* 102 (2): 720–49. See http://dx.doi.org/10.1257/aer.102.2.720.

Fujita, M., P. Krugman, and A. Venables. 1999. *The Spatial Economy: Cities, Regions, and International Trade.* Cambridge, MA: MIT Press.

Glaeser, E., J. Kolko, and A. Saiz. 2000. Technical report. Cambridge, MA: National Bureau of Economic Research.

Glinton, R., P. Scerri, and K. Sycara. 2010. "Exploiting Scale Invariant Dynamics for Efficient Information Propagation in Large Teams." In *Proceedings of the Ninth International Conference on Autonomous Agents and Multiagent Systems* in Toronto, Canada. Richland, SC: International Foundation for Autonomous Agents and Multiagent Systems. May 10–14.

Gneezy, U., S. Meier, and P. Rey-Biel. 2011. "When and Why Incentives (Don't) Work to Modify Behavior." *Journal of Economic Perspectives* 25 (4): 191–209.

Gomez-Rodriguez, M., J. Leskovec, and A. Krause. 2010. "Inferring Networks of Diffusion and Influence." In *Proceedings of the 16th ACM SIGKDD International Conference on Knowledge Discovery and Data Mining.* New York: ACM: 1019–28.

Gonzalez, M. C., C. A. Hidalgo, and A.-L. Barabási. 2008. "Understanding Individual Human Mobility Patterns." *Nature* 453 (June 5): 779–82; doi:10.1038/nature06958.

Granovetter, M. 1973. "The Strength of Weak Ties." *American Journal of Sociology* 78 (6): 1360–80.

———. 2005. "The Impact of Social Structure on Economic Outcomes." *Journal of Economic Perspectives* 19 (1): 33–50.

Granovetter, M., and R. Soong. 1983. "Threshold Models of Diffusion and Collective Behavior." *Journal of Mathematical Sociology* 9 (3): 165–79.

Gray, P. 2009. "Play as a Foundation for Hunter-Gatherer Social Existence." *American Journal of Play*, 1, 476–522.

Grund, T., C. Waloszek, and D. Helbing. 2013. "How Natural Selection Can Create Both Self- and Other-Regarding Preferences, and Networked Minds." *Scientific Reports* 3, no. 1480 (March 19); doi: 10.1038/srep01480.

Hägerstrand, T. 1952. "The Propagation of Innovation Waves." *Lund Studies in Geography: Series B, Human Geography*. no. 4. Sweden: Royal University of Lund.

———. 1957. "Migration and Area: Survey of a Sample of Swedish Migration Fields and Hypothetical Considerations of Their Genesis in Migration in Sweden, A Symposium." *Lund Studies in Geography: Series B, Human Geography*. no. 13. Sweden: Royal University of Lund.

Haidt, J. 2010. "The Emotional Dog and Its Rational Tail: A Social Intuitionist Approach to Moral Judgment." *Psychology Review* 108, no. 4: 814–34.

Hardin, G. 1968. "Tragedy of the Commons." *Science* 162, no. 3859 (December 13): 1243–48.

Hassin, R., J. Uleman, and J. Bargh, eds. 2005. *The New Unconscious*. Oxford Series in Social Cognition and Social Neuroscience. New York: Oxford University Press.

Helbing, D., W. Yu and H. Rauhut. 2011. "Self-organization and Emergence in Social Systems: Modeling the Coevolution of Social Environments and Cooperative Behavior." *Journal of Mathematical Sociology* 35 (1–3): 177–208.

Henrich, J., S. Heine, and A. Norenzayan. 2010. "The Weirdest People in the World?" *Behavioral and Brain Sciences* 33 (2–3): 61–83.

Hidalgo, C., B. Klinger, A.-L. Barabási, and R. Hausmann. 2007. "The Product Space Conditions the Development of Nations." *Science* 317, no. 5837 (July 27): 482–87.

Hidalgo, C. A., and C. Rodriguez-Sickert. 2008. "The Dynamics of a Mobile Phone Network." *Physica A: Statistical Mechanics and Its Applications* 387 (12): 3017–24; doi:10.1016/j.physa.2008.01.073.

Hong, L., and S. E. Page. 2004. "Groups of Diverse Problem Solvers Can Outperform Groups of High-Ability Problem Solvers." *Proceedings of the National Academy of Sciences* 101 (46): 16385–89; doi:10.1073/pnas.0403723101.

Iacoboni, M., and J. C. Mazziotta. 2007. "Mirror Neuron System: Basic Findings and Clinical Applications." *Annals of Neurology* 62 (3): 213–18.

Jacobs, J. 1961. *The Death and Life of Great American Cities*. New York: Random House.

Jaffe, A., M. Trajtenberg, and R. Henderson. 1993. "Geographic Localization of Knowledge Spillovers as Evidenced by Patent Citations." *Quarterly Journal of Economics* 108 (3): 577–98.

Kahneman, D. 2002. "Maps of Bounded Rationality," Nobel Prize Lecture. See http://www.nobelprize.org/nobel_prizes/economics/laureates/2002/kahneman-lecture.html.

———. 2011. *Thinking, Fast and Slow*. New York: Farrar, Straus and Giroux.

Kandel, E., and E. Lazear. 1992. "Peer Pressure and Partnerships." *Journal of Political Economy* 100 (4): 801–17.

Kelly, R. 1999. "How to Be a Star Engineer." *IEEE Spectrum* 36 (10): 51–58.

Kim, T. 2011. "Enhancing Distributed Collaboration Using Sociometric Feedback." PhD thesis, MIT.

Kim, T., A. Chang, L. Holland, and A. Pentland. 2008. Meeting Mediator: Enhancing Group Collaboration Using Sociometric Feedback." In *Proceedings of the 2008 ACM Conference on Computer Supported Cooperative Work*. New York: ACM: 457–66.

Kim, T., P. Hinds, and A. Pentland. 2011. "Awareness as an Antidote to Distance: Making Distributed Groups Cooperative and Consistent." In *Proceedings of the 2012 ACM Conference on Computer Supported Cooperative Work*. New York: ACM: 1237–46.

King, A. J., L. Cheng, S. D. Starke, and J. P. Myatt. 2012. "Is the True 'Wisdom of the Crowd' to Copy Successful Individuals?" *Biology Letters* 8, no. 2 (April 23): 197–200.

Kleinberg, J. 2013. "Analysis of Large-Scale Social and Information Networks." *Philosophical Transactions of the Royal Society* 371, no. 1987 (March): 20120378.

Krackhardt, D., and J. Hanson. 1993. "Informal Networks: The Company Behind the Chart." *Harvard Business Review* 71, no. 4 (July/August): 104–11.

Krause, S., R. James, J. J. Faria, G. D. Ruxton, and J. Krause. 2011. "Swarm Intelligence in Humans: Diversity Trumps Ability." *Animal Behaviour* 81 (5): 941–48; doi:10.1016/j.anbehav.2010.12.018.

Krugman, P. 1993. "On the Number and Location of Cities." *European Economic Review* 37 (2–3): 293–98.

Krumme, C. 2012. How Predictable: Modeling Rates of Change in Individuals and Populations. PhD thesis, MIT.

Krumme, C., M. Cebrian, G. Pickard, and A. Pentland. 2012. "Quantifying Social Influence in an Online Cultural Market." *PLoS ONE* 7 (5): e33785; doi:10.1371/journal.pone.0033785.

Krumme, C., A. Llorente, M. Cebrian, A. Pentland, and E. Moro. 2013.

"The Predictability of Consumer Visitation Patterns." *Scientific Reports* 3, no. 1645 (April 18); doi:10.1038/srep01645.

Lazer, D., and A. Friedman. 2007. "The Network Structure of Exploration and Exploitation." *Administrative Science Quarterly* 52 (4): 667–94.

Lazer, D., A. Pentland, L. Adamic, S. Aral, A.-L. Barabási, D. Brewer, N. Christakis, N. Contractor, J. Fowler, M. Gutmann, T. Jebara, G. King, M. Macy, D. Roy, and M. Van Alstyne. 2009. "Life in the Network: The Coming Age of Computational Social Science." *Science* 323, no. 5915 (February 6): 721–23.

Lee, R. B. 1988. "Reflections on Primitive Communism." In *Hunters and Gatherers, Vol. 1* ed. T. Ingold, D. Riches, and J. Woodburn, 252–68. Oxford, UK: Berg Publishers.

Lepri, B., A. Mani, A. Pentland, and F. Pianesi. 2009. "Honest Signals in the Recognition of Functional Relational Roles in Meetings." In *Proceedings of AAAI Spring Symposium on Behavior Modeling.* Stanford, CA.

Leskovec, J., K. Lang, A. Dasgupta, and M. Mahoney. 2009. "Community Structure in Large Networks: Natural Cluster Sizes and the Absence of Large Well-Defined Clusters." *Internet Mathematics* 6 (1): 29–123.

Lévi-Strauss, C. 1955. *Tristes Tropiques.* New York: Penguin Group (2012).

Liben-Nowell, D., J. Novak, R. Kumar, P. Raghavan, and A. Tomkins. 2005. "Geographic Routing in Social Networks." *Proceedings of the National Academy of Sciences* 102 (33): 11623–28.

Lim, M., R. Metzler, and Y. Bar-Yam. 2007. "Global Pattern Formation and Ethnic/Cultural Violence." *Science* 317, no. 5844 (September 14): 1540–44; doi: 10.1126/science.1142734.

Lima, A., M. De Domenico, V. Pejovic, and M. Musolesi. 2013. "Exploiting Cellular Data for Disease Containment and Information Campaign Strategies in Country-Wide Epidemics." See http://www.d4d.orange.com/home.

Lorenz, J., H. Rauhut, F. Schweitzer, and D. Helbing. 2011. "How Social Influence Can Undermine the Wisdom of Crowd Effect." *Proceedings of the National Academy of Sciences* 108 (22): 9020–25; doi:10.1073/pnas.1008636108.

Macy, M., and R. Willer. 2002. "From Factors to Actors: Computational Sociology and Agent-Based Modeling." *Annual Review of Sociology* 28: 143–66.

Madan, A., M. Cebrian, D. Lazer, and A. Pentland. 2010. "Social Sensing for Epidemiological Behavior Change." In *Proceedings of the 12th ACM International Conference on Ubiquitous Computing.* Ubicomp'10. Copenhagen: Denmark: ACM: 291–300; doi:10.1145/1864349.1864394.

Madan, A., M. Cebrian, S. Moturu, K. Farrahi, and A. Pentland. 2012.

"Sensing the 'Health State' of a Community." *IEEE Pervasive Computing* 11, no. 4 (October–December): 36–45.

Madan, A., K. Farrahi, D. G. Perez, and A. Pentland. 2011. "Pervasive Sensing to Model Political Opinions in Face-to-Face Networks." Lecture Notes in Computer Science. *Pervasive Computing*. 6696: 214–31.

Mani, A., C. M. Loock, I. Rahwan, and A. Pentland. 2013. "Fostering Peer Interaction to Save Energy." 2013 Behavior, Energy, and Climate Change Conference. Sacramento, CA. November 17.

Mani, A., C. M. Loock, I. Rahwan, T. Staake, E. Fleisch, and A. Pentland. 2012. "Fostering Peer Interaction to Save Energy." *International Conference on Information Systems (ICIS)*, Orlando, Florida, December 15–19.

Mani, A., A. Pentland, and A. Ozdalgar. 2010. "Existence of Stable Exclusive Bilateral Exchanges in Networks." See http://hd.media.mit .edu/tech-reports/TR-659.pdf.

Mani, A., I. Rahwan, and A. Pentland. 2013. "Inducing Peer Pressure to Promote Cooperation. *Scientific Reports* 3, no. 1735; doi:10.1038/ srep01735.

Marr, D. 1982. *Vision: A Computational Approach*. San Francisco: W. H. Freeman.

Marx, K. 1867. *Capital: Critique of Political Economy*. New York: Modern Library (1936).

Meltzoff, A. N. 1988. "The Human Infant as *Homo Imitans*." In *Social Learning* ed. T. R. Zentall and B. G. J. Galef. Hillsdale, NJ: Lawrence Erlbaum Associates. 319–41.

Milgram, S. 1974a. "The Experience of Living in Cities." In *Crowding and Behavior* ed. C. M. Loo. New York: MSS Information Corporation. 41–54.

———. 1974b. *Obedience to Authority: An Experimental View*. New York: Harper and Row.

Monge P. R., and N. Contractor. 2003. *Theories of Communication Networks*. New York: Oxford University Press.

Mucha, P., T. Richardson, K. Macon, M. Porter, and J. P. Onnela. 2010. "Community Structure in Time-Dependent, Multiscale, and Multiplex Networks." *Science* 328, no. 5980 (May 14): 876–78.

Myers, S., and J. Leskovec. 2010. "On the Convexity of Latent Social Network Inference." Neural Information Processing Systems conference. Vancouver, Canada. December 8.

Nagar, Y. 2012. "What Do You Think? The Structuring of an Online Community as a Collective-Sensemaking Process." In *Proceedings of the ACM 2012 Conference on Computer Supported Cooperative Work*. New York: ACM: 393–402.

Nguyen, T., and B. K. Szymanski. 2012. "Using Location-Based Social Networks to Validate Human Mobility and Relationships Models." In *Advances in Social Networks Analysis and Mining.* IEEE/ASONAM conference. Istanbul, Turkey (August 26): 1215–21. See http://arxiv.org/abs/1208.3653.

Nowak, M. 2006. "Five Rules for the Evolution of Cooperation." *Science* 314, no. 5805 (December 8): 1560–63; doi: 10.1126/science.1133755.

Olguín, D. O., B. Waber, T. Kim, A. Mohan, K. Ara, and A. Pentland. 2009. "Sensible Organizations: Technology and Methodology for Automatically Measuring Organizational Behavior." *IEEE Transactions on Systems, Man, and Cybernetics, Part B: Cybernetics,* 39 (1): 43–55. See http://web.media.mit.edu/~dolguin/Sensible_Organizations.pdf.

Onnela, J. P., S. Arbesman, M. Gonzalez, A.-L. Barabási, and N. Christakis. 2011. "Geographic Constraints on Social Network Groups." *PLoS ONE* 6 (4): e16939.

Onnela, J. P., J. Saramäki, J. Hyvönen, G. Szabó, D. Lazer, K. Kaski, J. Kertész, and A.-L. Barabási. 2007. "Structure and Tie Strengths in Mobile Communication Networks." *Proceedings of the National Academy of Sciences* 104 (18): 7332–36.

Ostrom, E. 1990. *Governing the Commons: The Evolution of Institutions for Collective Action.* Cambridge, UK: Cambridge University Press.

Pan, W., N. Aharony, and A. Pentland. 2011a. "Composite Social Network for Predicting Mobile Apps Installation." In *Proceedings of the Twenty-Fifth AAAI Conference on Artificial Intelligence.* Menlo Park, CA: AAAI Press. 821–27. See http://arxiv.org/abs/1106.0359.

———. 2011b. "Fortune Monitor or Fortune Teller: Understanding the Connection Between Interaction Patterns and Financial Status." In *Privacy, Security, Risk and Trust (PASSAT), 2011 IEEE Third International Conference on (IEEE).* Boston, MA (October 9–11): 200–7.

Pan, W., Y. Altshuler, and A. Pentland. 2012. "Decoding Social Influence and the Wisdom of the Crowd in Financial Trading Network." *Privacy, Security, Risk and Trust (PASSAT), 2012 International Conference on Social Computing,* Amsterdam, Netherlands. September 3–5; doi: 10.1109/SocialCom-PASSAT.2012.133.

Pan, W., W. Dong, M. Cebrian, T. Kim, J. Fowler, and A. Pentland. 2012. "Modeling Dynamical Influence in Human Interaction: Using Data to Make Better Inferences About Influence Within Social Systems." *Signal Processing.* 29 (2): 77–86.

Pan, W., G. Ghoshal, C. Krumme, M. Cebrian, and A. Pentland. 2013. "Urban Characteristics Attributable to Density-Driven Tie Formation." *Nature Communications* 4, no. 1961 (June 4); doi:10.1038/ncomms2961.

Papert, S., and I. Harel. 1991. "Situating Constructionism." *Constructionism*. 1–11.

Paridon, T., S. Carraher, and S. Carraher. 2006. "The Income Effect in Personal Shopping Value, Consumer Self-Confidence, and Information Sharing (Word-of-Mouth Communication) Research." *Academy of Marketing Studies* 10 (2): 107–24.

Pentland, A. 2008. *Honest Signals: How They Shape Our World*. Cambridge, MA: MIT Press.

———. 2009. "Reality Mining of Mobile Communications: Toward a New Deal on Data." In *The Global Information Technology Report 2008– 2009: Mobility in a Networked World*. eds. S. Dutta and I. Mia. Geneva: World Economic Forum. 75–80. See www.insead.edu/vl/gitr/ wef/main/fullreport/files/Chap1/1.6.pdf.

———. 2010a. "To Signal Is Human." *American Scientist* 98 (3): 204–10.

———. 2010b. "We Can Measure the Power of Charisma." *Harvard Business Review* 88, no. 1 (January–February): 34–35.

———. 2011. "Signals and Speech." In *Twelfth Annual Conference of the International Speech Communication Association*. Florence, Italy (August 28–31).

———. 2012a. "Society's Nervous System: Building Effective Government, Energy, and Public Health Systems." *IEEE Computer* 45 (1): 31–38.

———. 2012b. "The New Science of Building Great Teams." *Harvard Business Review* 90, no. 4 (April): 60–69. See http://www.ibd corporation.net/images/buildingteams.pdf.

———. 2012c. "Reinventing Society in the Wake of Big Data: A Conversation with Alex (Sandy) Pentland." Edge.org (August 30). See http://www.edge.org/conversation/reinventing-society-in-the-wake-of -big-data.

———. 2013a. "Strength in Numbers." To appear in *Scientific American*, October 2013.

———. 2013g. "Beyond the Echo Chamber." *Harvard Business Review*. November 2013.

Pentland, A., D. Lazer, D. Brewer, and T. Heibeck. 2009. "Improving Public Health and Medicine by Use of Reality Mining." In *Studies in Health Technology Informatics*, 149. Amsterdam, Netherlands: IOS Press. 93–102.

Pickard, G., W. Pan, I. Rahwan, M. Cebrian, R. Crane, A. Madan, and A. Pentland. 2011. "Time-Critical Social Mobilization." *Science* 334, no. 6055 (October 28): 509–12; doi: 10.1126/science.1205869.

Pink, D. 2009. *Drive: The Surprising Truth About What Motivates Us*. New York: Penguin.

Pong, S., and D. Ju. 2000. "The Effects of Change in Family Structure and

Income on Dropping Out of Middle and High School." *Journal of Family Issues* 21 (2): 147–69.

Prelec, D. 2004. "A Bayesian Truth Serum for Subjective Data." *Science* 306, no. 5695 (October 15): 462–66.

Putnam, R. 1995. "Bowling Alone: America's Declining Social Capital." *Journal of Democracy* 6 (1): 65–78.

Rand, D. G., A. Dreber, T. Ellingsen, D. Fudenberg, and M. A. Nowak. 2009. "Positive Interactions Promote Public Cooperation." *Science* 325, no. 5945 (September 4): 1272–75.

Reagans, R., and E. Zuckerman. 2001. "Networks, Diversity, and Productivity: The Social Capital of Corporate R&D Teams." *Organization Science* 12 (4): 502–17.

Rendell, L., R. Boyd, D. Cownden, M. Enquist, K. Eriksson, M. W. Feldman, L. Fogarty, S. Ghirlanda, T. Lillicrap, and K. N. Laland. 2010. "Why Copy Others? Insights from the Social Learning Strategies Tournament." *Science* 328, no. 5975 (April 9): 208-13.

Rutherford, A., M. Cebrian, S. Dsouza, E. Moro, A. Pentland, and I. Rahwan. 2013. "Limits of Social Mobilization." *Proceedings of the National Academy of Sciences* 110 (16): 6281–86.

Saavedraa, S., K. Hagerty, and B. Uzzi. 2011. "Synchronicity, Instant Messaging, and Performance Among Financial Traders." *Proceeding of the National Academy of Sciences* 108 (13): 5296–301.

Salamone, F. A. 1997. *The Yanomami and Their Interpreters: Fierce People or Fierce Interpreters?* Lanham, MD: University Press of America.

Salganik, M., P. Dodd, and D. Watts. 2006. "Experimental Study of Inequality and Unpredictability in an Artificial Cultural Market. *Science* 311, no. 5762 (February 10): 854–56.

Sartre, J.-P. 1943. *Being and Nothingness / L'étre et le néant*. New York: Philosophical Library (1956).

Schneider, M. J. 2010. *Introduction to Public Health*. Sudbury, MA: Jones and Bartlett.

Schwartz, P. 2003. "Property, Privacy, and Personal Data." *Harvard Law Review* 117: 2056.

Shmueli, E., Y. Altshuler, and A. Pentland. 2013. "Temporal Percolation in Scale-Free Networks." *International School and Conference on Network Science (NetSci)*. Copenhagen, Denmark, June 5–6.

Sigmund, K., H. De Silva, A. Traulsen, and C. Hauert. 2010. "Social Learning Promotes Institutions for Governing the Commons." *Nature* 466 (August 12): 861–63.

Simon, H. 1978. "Rational Decision Making in Business Organizations," Nobel Prize in Economic Sciences lecture. See http://www.nobelprize .org/nobel_prizes/economics/laureates/1978/simon-lecture.html.

Singh, V., E. Shmueli, and A. Pentland. "Channels of Communication;" in preparation.

Slemrod, J. 1990. "Optimal Taxation and Optimal Tax Systems." *Journal of Economic Perspectives* 4 (1): 157–78.

Smith, A. 1937. *The Wealth of Nations.* New York: Modern Library, 740.

———. 2009. *Theory of Moral Sentiments.* New York: Penguin Classics.

Smith, C., A. Mashadi, and L. Capra. 2013. "Ubiquitous Sensing for Mapping Poverty in Developing Countries." See http://www.d4d.orange.com/home.

Smith, C., D. Quercia, and L. Capra. 2013. "Finger on the Pulse: Identifying Deprivation Using Transit Flow Analysis." In *Proceedings of the 2013 Conference on Computer Supported Cooperative Work.* New York: ACM: 683–92; doi: 10.1145/2441776.2441852.

Snijders, T. A. B. 2001. "The Statistical Evaluation of Social Network Dynamics." *Sociological Methodology* 31 (1): 361–95.

Stewart, K. J., and A. H. Harcourt. 1994. "Gorilla Vocalizations During Rest Periods: Signals of Impending Departure." *Behaviour* 130 (1–2): 29–40.

Sueur, C., A. King, M. Pele, and O. Petit. 2012. "Fast and Accurate Decisions as a Result of Scale-Free Network Properties in Two Primate Species." *Proceedings of the Complex System Society* (January).

Surowiecki, J. 2004. *The Wisdom of Crowds: Why the Many Are Smarter Than the Few and How Collective Wisdom Shapes Business, Economies, Societies and Nations.* London: Little Brown.

Sweeney, L. 2002. "k-anonymity: A Model for Protecting Privacy." *International Journal of Uncertainty, Fuzziness and Knowledge-Based Systems* 10 (05): 557–70.

Tetlock, P. E. 2005. *Expert Political Opinion: How Good Is It? How Can We Know?* Princeton, NJ: Princeton University Press.

Tett, G. "Markets Insight: Wake Up to the #Twitter Effect on Markets." *Financial Times,* April 18, 2013. See http://www.physiciansmoneydigest.com/personal-finance/Wake-up-to-the-Twitter-Effect-on-Markets-FT.

Thomas, E. M. 2006. *The Old Way: A Story of the First People.* New York: Farrar, Straus and Giroux.

Tran, L., M. Cebrian, C. Krumme, and A. Pentland. 2011. "Social Distance Drives the Convergence of Preferences in an Online Music-Sharing Network." *Privacy, Security, Risk and Trust (PASSAT), 2011 IEEE Third International Conference on Social Computing.* Boston, MA, October 9–11.

Tripathi, P. 2011. Predicting Creativity in the Wild. PhD thesis, Arizona State University.

Tripathi, P., and W. Burleson. 2012. "Predicting Creativity in the Wild:

Experience Sample and Sociometric Modeling of Teams." In *Proceedings of the ACM 2012 Conference on Computer Supported Cooperative Work*. Seattle, WA (February 11-15). New York: ACM: 1203–12.

Uzzi, B. 1997. "Social Structure and Competition in Interfirm Networks: The Paradox of Embeddedness." *Administrative Science Quarterly* 42 (1): 35–67.

Waber, B. 2013. *People Analytics: How Social Sensing Technology Will Transform Business and What It Tells Us About the Future of Work*. Upper Saddle River, NJ: FT Press.

Watts, D. J., and P. S. Dodds. 2007. "Influentials, Networks, and Public Opinion Formation." *Journal of Consumer Research* 34 (4): 441–58.

Weber, M. 1946. "Class, Status, Party." In *From Max Weber: Essays in Sociology*, eds. H. Gerth and C. Wright Mills. Abingdon, UK: Routledge. 180–95.

Wellman, B. 2001. "Physical Place and Cyberplace: The Rise of Personalized Networking." *International Journal of Urban and Regional Research* 25 (2): 227–52.

White, H. 2002. *Markets from Networks: Socioeconomic Models of Production*. Princeton, NJ: Princeton University Press.

Wirth, L. 1938. "Urbanism as a Way of Life." *American Journal of Sociology* 98 no. 1 (July): 1–24.

Woolley, A., C. Chabris, A. Pentland, N. Hashmi, and T. Malone. 2010. "Evidence for a Collective Intelligence Factor in the Performance of Human Groups." *Science* 330, no. 6004 (October 29): 686–88. doi: 10.1126/science.1193147.

World Economic Forum. 2011. Personal data: The emergence of a new asset class. See http://www3.weforum.org/docs/WEF_ITTC_PersonalDataNewAsset_Report_2011.pdf.

Wu, L., B. Waber, S. Aral, E. Brynjolfsson, and A. Pentland. 2008. Mining face-to-face interaction networks using sociometric badges: Predicting productivity in an IT configuration task. Available at Social Science Research Network (SSRN) working papers series 1130251 (May 7).

Wyatt, D., T. Choudhury, J. Bilmes, and J. Kitts. 2011. "Inferring Colocation and Conversation Networks from Privacy-Sensitive Audio with Implications for Computational Social Science." *ACM Transactions on Intelligent Systems and Technology (TIST)* 2 no. 1 (January): 7.

Yamamoto, S., T. Humle, and M. Tanaka. 2013. "Basis for Cumulative Cultural Evolution in Chimpanzees: Social Learning of a More Efficient Tool-Use Technique." *PLoS ONE* 8 (1): e55768; doi:10.1371/journal.pone.0055768.

Zimbardo, P. 2007. *The Lucifer Effect: Understanding How Good People Turn Evil.* New York: Random House.

Zipf, G. K. 1946. The $P_1 P_2 / D$ hypothesis: On the inter-city movement of persons. *American Sociological Review* 11, no. 6 (December): 677–86.

———. 1949. *Human Behavior and the Principle of Least Effort.* Cambridge, MA: Addison-Wesley Press. See http://en.wikipedia.org/wiki/Zipf's_law.

Index